how to rebuild your FORD V8 351C-351M-400-429-460

by Tom Monroe
**Registered Professional Engineer;
Member, Society of Automotive Engineers**

Chapter 1
DO YOU NEED TO REBUILD? 3

Chapter 2
ENGINE REMOVAL 12

Chapter 3
PARTS IDENTIFICATION AND INTERCHANGE 22

Chapter 4
TEARDOWN 39

Chapter 5
INSPECTING AND RECONDITIONING THE SHORTBLOCK ... 51

Chapter 6
HEAD RECONDITIONING AND ASSEMBLY 74

Chapter 7
ENGINE ASSEMBLY 91

Chapter 8
CARBURETOR AND DISTRIBUTOR REBUILD 124

Chapter 9
ENGINE INSTALLATION 137

Chapter 10
TUNEUP 157

ANOTHER FACT-FILLED HP AUTO BOOK

NOTICE: The information contained in this book is true and complete to the best of our knowledge. All recommendations on parts and procedures are made without any guarantee on the part of the author or **HPBooks**. Because the quality of parts, materials and methods are beyond our control, author and publisher disclaim all liability incurred in connection with the use of this information.

The cooperation of Ford Motor Company is gratefully acknowledged. However, this publication is a wholly independent production of H.P.Books (Fisher Publishing, Inc.).

Publisher: Tom Monroe; Editor: Bill Fisher; Art Director: Don Burton; Book Design and Assembly: Tom Jakeway; Typography: Cindy Coatsworth, Joanne Porter, Kris Spitler; Drawings and photos: Tom Monroe; Cover photo: Bill Keller.

Published by **H.P.Books**, P.O.Box 5367, Tucson, AZ 85703 602/888-2150
ISBN 0-89586-036-8
Library of Congress Catalog Card Number, 80-80171 ©1980 Fisher Publishing, Inc.
Printed in U.S.A.

Introduction

This book deals with rebuilding two separate families or series of Ford V8s. Unfortunately they cannot be simply defined as "big-block" or "small-block" engines, but use designations you may be unfamiliar with—"335 Series" and "385 Series." Engines in the **335 Series** are: 351C, 351C Boss, 351C HO, 351M and 400. **385 Series** engines include: 429, 429CJ, 429SCJ, 429 Police, 460 and 460 Police.

A couple of obvious questions arise: Why don't the series numbers correlate to any of the engine displacements? And, what do the letter suffixes attached to the displacements mean. Series numbers are actually code numbers used by engineering to distinguish one new engine project from another. It doesn't necessarily have anything to do with the engine's displacement. As for the suffixes, they distinguish engines having the same displacements, with both major and minor design differences. The 351C started it where **C** represents "Cleveland"—the Ford engine plant in which it was produced. This sets it apart from the 351W—**W** indicating the location of its engine plant—Windsor, Canada. Although these two engines have some design similarities, there is no practical parts interchangeability. **M** in 351M doesn't mean anything. It's just there to set it apart from the 351W and 351C engines even though many have taken the M to mean *modified* or *Michigan*. Other suffixes represent performance versions of a base engine; **HO** is for High Output, **CJ** stands for Cobra Jet and **SCJ** is Super Cobra Jet.

Ford's 335 and 385 engine series were conceived during the peak of the great horsepower race era in the mid-60s. As a result, they are large-displacement engines, particularly the 429 and 460. Series 385 engines were originally intended for use in large passenger cars, however they have been forced out of the passenger-car application in favor of smaller displacement engines due to Federal mandates and the fuel shortage. As a result, most of these engines are now found in trucks.

One similarity shared by the 335 and 385 Series engines is their cylinder-head designs. They use canted valves and large intake and exhaust ports for maximum *breathing*. A *canted* valve is angled from its vertical position as you would view the cylinder head from its side. This tends to provide better "breathing," that is, a port of the same size flows more air with a canted valve than it would with a straight up, or vertical valve. An engine of a given displacement that can breathe in more air and fuel through its intake ports and discharge more burned mixture through its exhaust ports will produce more power. If done properly, there will be no concurrent loss in fuel economy. This was accomplished with the 335 and 385 engines. As the classic example, 351Cs are very well known by their competition performances on the NASCAR "high banks" as well as in Pro Stock drag racing for their power-producing capability. On the other hand, the 429s and 460s used in everyday applications are known for getting gas mileage that engines two-thirds their size should get—without any sacrifice in power.

Let's get on with what this book is about—keeping your engine around longer by restoring it to peak mechanical condition. Rebuilding an engine seems like an ominous undertaking to the person who's never done it. And ominous it is, particularly when you consider the number of components in an engine and the decisions that have to be made during a rebuild. However, the difficulty of doing a rebuild is reduced in direct proportion to the information you have about your engine. The more you have, the easier the job. The less you have, the more difficult it is—to the point of being impossible.

While preparing this book I rebuilt four engines and photographed every step—determining whether the engine needed rebuilding, or what it needed, removing it from the vehicle, tearing it down, inspecting it, reconditioning the parts and reassembling them into complete engines and finally reinstalling the engines and breaking them in.

I tried to leave nothing to your imagination. This book has more information on how to tear an engine down and reassemble it than any other book I know about. It includes information to make you the expert on your engine; tells what you need to do the job. I discuss what you can and can't "get away with," to keep you from being victimized by well-meaning but inaccurate information. The information I include in words *and* pictures is a product of my experience and knowledge and that of experts who make their living rebuilding engines, tuning them and supplying new and reconditioned parts for Fords.

A final word before getting started. Work safely and use the right tools for the specific job. Take nothing for granted—*check everything*—and be alert. A bearing cap installed backwards or in the wrong location will spell the difference between a successful engine rebuild and at least a $200 disaster. With those points in mind, let's get into your engine.

1 | Do You Need To Rebuild?

Before tearing down your engine, you should determine to what extent it needs rebuilding, if at all. Your engine must have seen thousands of miles, be noticeably down on power, getting poor mileage or consuming excessive amounts of oil—otherwise you shouldn't be considering a rebuild.

Just because an engine has accumulated a lot of miles, it doesn't necessarily need rebuilding. A Ford V8 can exceed the 100,000-mile mark and still provide excellent service. For example, an engine which powers a truck used in a rock quarry naturally inhales more dust than one in the typical family sedan. The result is more internal wear per hour of use—particularly the piston-ring and ring-land wear. An engine that has been frequently over-revved will also wear excessively, and may possibly have internal damage such as bent pushrods.

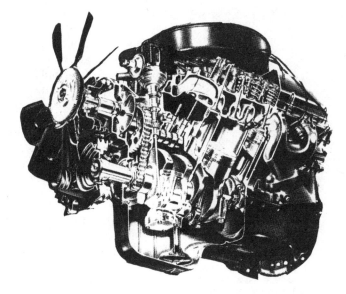

Internal view of the 429/460. This engine started its life in luxury, in Lincolns and Thunderbirds, eventually found its way into Torinos, Montegos, Mustangs and Cougars as a performance engine, and ended up as a workhorse in trucks. Photo courtesy Ford.

EXCESSIVE OIL CONSUMPTION

How much oil an engine uses is the most common "yardstick" used to judge engine condition—and rightly so. Oil consumption is largely determined by clearances between moving parts. As an engine wears, these clearances increase, resulting in increased oil consumption *and* oil-pressure loss.

What should be considered excessive oil consumption? I think any engine using oil at the rate of one quart in less than 1,000 miles needs attention. If a quart is lost in 500 miles or less, the condition is serious.

Other than the obvious gasket or seal leak which shows up as large oil spots on your driveway or garage floor, the two major causes of oil consumption are worn or broken piston rings and worn valve guides. These let oil escape past the pistons or intake-valve stems into the combustion chamber, or past the exhaust-valve stems into the exhaust creating an ominous puff of blue smoke from the exhaust pipe when starting an engine or when applying power after decelerating.

Piston Rings—Three rings per piston are used. The two top rings seal compression and combustion pressures. The second compression ring also keeps oil from getting into the combustion chamber. Oil control is the job of the bottom ring—the oil ring. It doesn't completely seal the cylinder from the crankcase, otherwise the compression rings and piston wouldn't receive any lubrication and would seize or scuff the cylinder bore and wear away the rings and ring lands.

If oil can get past the rings into the combustion chamber, compression and combustion pressures can blow past the rings into the crankcase, called *blowby*. Blowby is recycled back into the combustion chamber via the PCV (positive crankcase ventilation) system and is burned unless it is excessive. It then appears as oil deposits on the top of the pistons, in the combustion chamber, on the valves, spark plugs and throughout the exhaust system.

Valve Guides—A worn valve guide allows oil to get past the valve stem and into the combustion chamber or exhaust port, depending on whether it is an intake or exhaust valve. Because more oil must be used for lubrication, there should always be some controlled oil loss through the valve guides.

LOSS IN PERFORMANCE?

When referring to performance, I mean fuel economy as well as power. As an engine wears, its performance suffers. A change in fuel economy or oil consumption is easier to monitor than the power output of your engine because you can

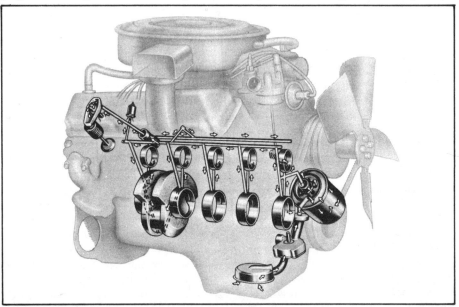

429/460 oiling system is typical of all Ford V8s. Main oil galleries fed by the oil pump via the oil filter, supply oil pressure to the lifters, main, connecting-rod and camshaft bearing journals. Photo courtesy Ford.

Monitoring engine functions under simulated road-load conditions is a good way to determine whether your engine is "tired" and needs rebuilding.

measure it. Judging power loss can be done with a chassis dynamometer. Using the "seat-of-the-pants" method is inaccurate because the wearing-out process and accompanying power loss are too gradual.

If you suspect your engine's performance is suffering, but oil consumption is normal, give your engine a thorough tuneup. Choose a reputable tuneup shop with a chassis dynamometer. It measures power at the drive wheels while critical engine functions are being monitored. You can compare horsepower readings before and after the tuneup. Just make sure that the tuneup shop you take your car to can give you horsepower readings. Many don't, or can't. If the tuneup cures the problem, relax and read the rest of the book for entertainment just to see all the fun you missed. If it doesn't, you'll have to do some further investigation.

CAUSE OF POOR PERFORMANCE

Let's review some possible causes of performance loss. The first suspect is piston-ring and cylinder-bore wear. These cause increased oil consumption. If blow-by is excessive, there will be an accompanying loss of compression and combustion pressure. The engine produces less power and gets worse fuel economy. If your engine is "hurting" in both the oil-consumption and performance departments, chances are the problem is with the rings and cylinder bores. If this is not the case, look further.

If your engine is giving poor gas mileage and power, but is OK on oil consumption, suspect the valve train. This assumes the carburetion and ignition systems are in good order. Problems can include a blown head gasket, burned exhaust valves, worn camshaft lobes and lifters, and carbon buildup in the combustion chamber.

Blown Head Gasket—A blown head gasket causes compression and combustion pressure to drop much in the same manner as bad rings and cylinder bores, only worse. Pressure lost past a gasket goes into the cooling system or an adjacent cylinder. If it leaks into the cooling system, only one cylinder will be affected, making if difficult to detect the loss in power and gas mileage.

A gasket blown or leaking *between* cylinder bores affects two cylinders, consequently power loss is easier to detect. One thing is sure, if cylinder pressure is getting into your cooling system, your engine will eventually overheat. The cooling system will be overpressurized from cylinder pressures, forcing coolant out of the radiator.

Burned Exhaust Valves—A burned exhaust valve can't seal its combustion chamber. Therefore, the cylinder with the bad valve will produce less compression and power. This occurs more often in emission-controlled engines which operate at higher temperatures for reduced emissions. The valves run at higher temperatures and are more susceptible to burning.

Worn Camshaft Lobes and Lifters—Worn camshaft lobes and lifters almost always occur together. This problem does not affect an engine's oil consumption, but really reduces its performance. The more worn lobes, the more performance is affected. Also, the metal particles being circulated through an engine's oiling system don't help any.

As a cam lobe and lifter wear, the lift of the valve they operate becomes less. If it is an intake valve, a smaller fuel charge enters the combustion chamber, causing reduced performance. The same thing happens with the exhaust valve, but in a roundabout way. If valve lift is reduced, all exhaust gases cannot leave the combustion chamber, consequently the new fuel charge drawn in when the intake valve opens will be diluted by exhaust gases, reducing power output.

Carbon Deposits—Carbon deposits are not a direct result of how many miles are on an engine or its age, but are caused by how the vehicle is used. A vehicle operated on the open highway won't experience excessive carbon buildup, assuming the carburetion is near right. However, one that is used for puttering around town at 30 MPH or so may develop the problem. Carbon deposits don't require that an engine be rebuilt to remedy the problem.

But, because its symptoms can fool you, I'll discuss how carbon can affect an engine, and how you can remedy the problem.

Carbon buildup takes up room in the combustion chamber, raising the compression ratio. As a result, *detonation* problems may develop, usually called *pinging* or *knocking*. This is caused by the fuel charge *exploding* from compression rather than *burning smoothly*. Higher loads are imposed on an engine by detonation, and this can cause serious damage. Damage can range from deformed main-bearing caps to broken piston rings, and even pistons. Detonation also blows head gaskets. *Preignition* may also occur when the carbon gets hot and acts like a two-cycle model-airplane-engine glow plug, igniting the fuel charge prematurely. This potentially serious problem can melt pistons, and break piston rings. Carbon deposits also cause *dieseling*, or continuing to run after the ignition is turned off—sometimes turning the crank in the opposite direction.

Detonation, preignition and dieseling don't necessarily hurt an engine's performance, but the damage that can result, particularly from detonation and preignition, should concern you.

Carbon deposits hurt performance in a couple of ways. Buildup around the valves reduces flow to and from a combustion chamber, thereby hurting power output. And, pieces of carbon can break loose and go out the exhaust harmlessly, or end up on the piston top, between the exhaust valve and its seat, or between the spark-plug electrodes. Carbon on top of a piston can reduce the clearance between the piston and the head. The engine can develop a knock, giving the impression a rod bearing is bad when it isn't. I don't know anything this can hurt except your piece of mind.

Carbon between an exhaust valve and seal prevents the valve from fully closing, thus that cylinder won't be producing full power due to lost pressure. Also, on the power stroke the hot fuel charge will escape past the valve and seal, overheating the valve with a good possibility of burning it.

As for carbon between a spark plug's electrodes, the plug will be shorted, preventing it from igniting the fuel charge. A misfiring cylinder results.

What usually causes carbon to break off and create the troubles I've just discussed is someone taking their car out and "blowing the carbon out" or taking advantage of an additive sale, then going home and dumping the "instant overhaul" solution down the carburetor. This stuff *really* works, loosening the carbon which then causes these problems. Don't try the "cure-all" approach to rid your engine of carbon. Use the methods I discuss in the block and cylinder-head reconditioning chapter. Also, carbon is an effect rather than a cause. It results from an excessively rich fuel mixture, oil getting into the combustion chamber past the piston or intake-valve stem, very slow driving or idling for extended periods. If carbon-buildup causes are cured, the deposits gradually burn away, negating any need to tear your engine down to remove the deposits mechanically.

DIAGNOSIS

Now that I've discussed the types of internal problems you may encounter with your engine and how each may affect its operation, let's look at how to diagnose these problems without tearing your engine down. On the other hand, your engine may not have any specific problems, but you do want to determine if it's time to rebuild.

Internal Noises—Perhaps your engine has noises of impending disaster coming from its innards. They may or may not be accompanied by an increase in fuel or oil consumption or a reduction in power. Generally, if the noise is at engine speed—once for every revolution of the crank—the problem is in the *bottom end*. Causes are a broken piston ring, worn connecting-rod bearing or a worn piston-pin bore.

A noise at half engine speed or camshaft speed, is probably in the valve train, with one exception. *Piston slap* occurs only on the power stroke, consequently it is also at half speed. If the noise is coming from the valve train, it could be due to excessive *lash* or clearance in the valve train caused by a collapsed hydraulic lifter, too much valve clearance or a bent pushrod. To help in determining the speed of the noise, hook up your timing light and watch the light while listening to your engine. If the light flashes in time with the noise, it's at half engine speed.

To assist in listening to what's going on inside your engine, use a long screwdriver and press its tip against the block or cylinder head close to the area where you suspect the noise to be coming from. Press your head, or skull just below your ear against the handle. This will amplify engine noises. Just make sure you place the screwdriver against a solid part of the engine to get the best noise transmittal. Don't put the end of the screwdriver against the valve cover when listening to

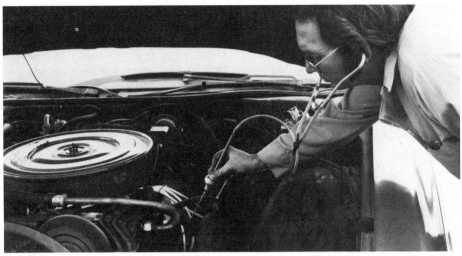

Noise-related engine problems can be located with an automotive stethoscope. You must be able to distinguish between normal engine noises and those which are not.

valve-train noises. Put it against the head or a valve-cover bolt. The cork gasket between the cover and the head and the large air space under the valve cover muffles much of the noise.

What should the different noises sound like? Let's start with the bottom end. A broken piston ring makes a chattering or rattling noise which is more prevalent during acceleration. A dull or hollow sound is usually caused by piston slap, or the piston wobbling and *slapping* against the side of its bore due to excess clearance between the piston and bore. A collapsed piston skirt causes a similar, but louder noise. Slap resulting from excess clearance will be loudest when the engine is cold. It decreases as the engine warms up and the piston *grows* to reduce piston-to-bore clearance. You can check for piston slap by retarding the spark. Loosen your distributors's hold-down bolt and rotate it counterclockwise about 5 degrees. This will retard the spark and should reduce any noise due to piston slap. Use your timing light to reset your timing after you've made this check.

A light knocking or pounding noise that's not related to detonation or preignition is probably excess connecting-rod-bearing-to-journal clearance. Simply put, the bearing is worn out. Finally, a light tapping noise can indicate excess piston-pin clearance in the piston.

To confirm and pinpoint a bottom-end noise-related problem, disconnect the spark-plug leads one at a time, then run your engine and listen for the noise to change or go away. What happens is the power-stroke is eliminated from the cylinder with the disconnected spark-plug lead, thus unloading its piston and connecting rod. So, if the noise is piston or rod-related, it will be greatly reduced or eliminated when you have the plug wire disconnected from the problem cylinder. If your car is equipped with a solid-state ignition, *always ground the lead you disconnect*. Otherwise you risk damaging your expensive ignition system.

A sharp clacking or rapping noise indicates your engine probably has a collapsed hydraulic lifter. If the noise is a light clicking, it is probably excess clearance in one of the valve mechanisms. This assumes your engine is not equipped with solid lifters which click normally—unless the clicking is excessive. Then the problem will also be excess clearance, or lash. If

---

SOLID-STATE IGNITION SYSTEMS
Many solid-state (electronic) ignition systems generate very high secondary-voltage peaks when a spark-plug lead is unloaded by disconnecting it from its spark plug, or by disconnecting the coil-to-distributor lead while the engine is being cranked or is running. The resulting secondary voltage surge of up to 60,000 volts can damage a coil internally, or pierce plug-wire insulation or a distributor cap as the high voltage seeks a ground. So, if you remove any secondary lead with the ignition on and the engine turning, ground the lead to the engine with a jumper wire. Ford has installed solid-state (breakerless) ignitions since 1974.

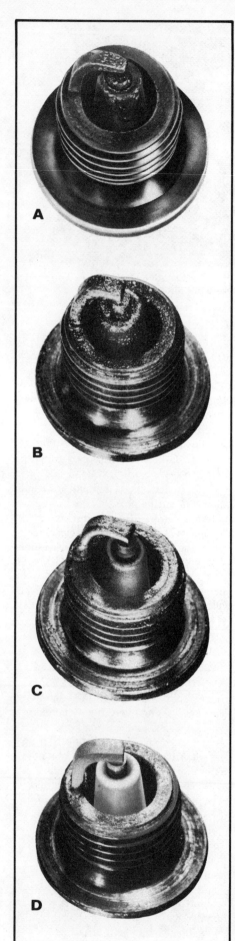

this is the problem, a simple valve adjustment may correct things.

To check for proper lash, remove the valve cover from the noisy side of the engine and insert a feeler gage between each rocker arm and its valve stem one at a time. When you get to the noisy one, the gage will take up lash and reduce the noise. In this instance, a simple adjustment may correct the problem if your rocker arms are adjustable. If there is still a lot more clearance when the feeler gage is between the rocker and the valve-stem tip, suspect that the lifter is malfunctioning or the pushrod is bent, particularly if the engine was over-revved.

Spark Plugs Tell a Story— Prior to testing, an easy way to diagnose your engine is to "look" into the combustion chambers by removing the spark plugs and inspecting them. Each plug has a story to tell about the cylinder it came out of, so remember to keep them in order.

The main thing to look for is a wet-black deposit on the inside of the threaded portion of the plug, and on the plug insulator. This is caused by oil getting past the rings or the intake-valve stems. Oil loss through exhaust-valve guides doesn't enter the combustion chamber, and won't show on the plugs.

Black, dry and fluffy deposits are carbon caused by an over-rich fuel mixture, excessive idling or driving at sustained low speeds without much load on the engine. All the plugs should appear about the same in this case. The fuel mixture will have to be corrected by tuning, however carbon buildup caused by the way a vehicle is operated can be partially corrected by installing hotter plugs. This keeps the carbon burned off the plugs so they won't foul, but it doesn't eliminate carbon buildup on the intake ports or on the back of the intake valves.

Cranking Vacuum Test—An internal-combustion engine is a specialized air pump, so how it sucks air, or *pulls a vacuum* is an indicator of its mechanical soundness, or how your engine's cylinders are sealing relative to one another. You'll need a vacuum gage and a remote starter, or a friend to operate the starter switch while

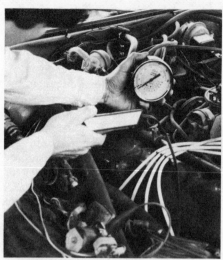

Vacuum gage being used to check the pumping ability of this engine. A fluctuating needle indicates problems in at least one cylinder.

you watch the vacuum gage. The remote starter lets you control the starter from under the hood.

When doing this test, connect the vacuum gage to the intake manifold *after* you've warmed up your engine. Disable the ignition system so the engine can't be started. Disconnect the high-tension, or distributor-to-coil lead so the engine won't start when it is being cranked. With electronic systems, disconnect the distributor-to-amplifier lead. With your eye on the vacuum gage, crank the engine. If the gage indicates a steady vacuum reading after it has stabilized, all eight cylinders are sealing the same in relation to one another. If the needle fluctuates, indicating a pulsating vacuum, one or more of the cylinders has a problem, assuming your starter is cranking steadily. It could be valve timing due to incorrect adjustment, a worn camshaft lobe or collapsed valve lifter, a leaking valve, worn cylinder bore or piston rings, a broken piston ring or a leaky head gasket. If this indicates a bad cylinder/s, the next two tests pinpoint which one it is.

Power-Balance Test— A power-balance test determines if all of your engine's

"Reading" your spark plugs is a good troubleshooting technique. Worn-out plugs as in A are easily recognizable by eroded electrodes and pitted insulator. Replace them and your engine's performance will improve instantly. Oil-coated and fouled plug B indicates internal engine wear: piston rings, cylinder bores and valve guides. Carboned plug C coated with dry, black and fluffy deposits is usually caused by carburetion problems or driving habits. Normal spark plug D has a brown to greyish-tan appearance with some electrode wear indicated by slightly radiused electrode edges. When inspecting your plugs, keep track of the cylinder each belongs to. Photos courtesy Champion Spark Plug Company.

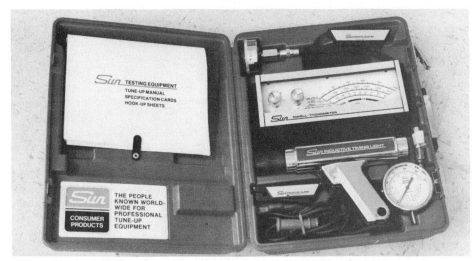

A complete, high-quality tune-up kit eventually pays for itself. Sun kit includes a compression tester, tach/dwell meter, timing light, vacuum gage and a remote starter switch. Low-RPM scale on the tach/dwell meter has the accuracy required to perform power-balance testing.

 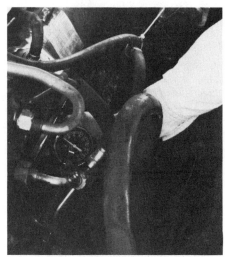

A thread-in compression gage is much easier to use than the type which has to be held into place. Regardless of the type of gage you're using, the lowest pressures should be not less than 75 percent of the highest pressure.

cylinders are contributing the same amount of power. The method is to fast-idle your engine, then disconnect the spark plugs one-by-one and monitor the RPM drop. RPM change indicates how much each cylinder is contributing to overall engine power. The less it drops, the less power that cylinder is contributing. You'll need a tachometer—one that is very accurate such as found on a good *dwell tachometer*. Slide the rubber boots back over their spark-plug leads at the distributor cap if you have the early-style leads. Otherwise, you can bend eight pieces of wire so they'll protrude out the bottom of the boots. They can be clipped onto with a jumper. This will disable each plug individually without having to disconnect the leads. Use a jumper wire grounded to the engine at one end and stripped of its insulation at the other so the jumper will fit beside each lead in the cap. Just be careful you don't end up being the ground. Using insulated pliers to handle the jumper will help keep you from being jolted by a good shot of high-voltage electricity.

With your engine warmed up and the tach connected, start your engine and set its idle speed to about 1,000 RPM. Disable cylinder 1 spark plug. Record engine-RPM drop after it stabilizes at the lower level. Reconnect the lead and let the RPM return to normal, then disable the next plug in the firing order. If all drops are within 20 RPM, no one cylinder is much worse than its brothers. Any drop which differs 40 RPM or more indicates that cylinder has a problem. Note the cylinders with the least drop, and concentrate on them during compression testing.

Compression Testing—Comparative compression testing of an engine's cylinders gives you an idea of the condition of its piston rings and cylinder bores. You'll need a compression tester, a note pad and a friend or a remote starter switch.

Run your engine until it is up to operating temperature. Shut the engine off, disconnect the ignition wires and remove all the spark plugs, being careful not to get burned—everything is **HOT**. Now for the test. Prop the throttle open and make sure the choke is also open. Insert a tester in cylinder 1 spark-plug hole. A screw-in-type tester makes this whole procedure a lot easier. If yours has a rubber cone on the end, insert it into the spark-plug hole while holding it

DIAGNOSIS 7

If a cylinder is down on pressure, give it a few shots of oil, then recheck its pressure after waiting a few minutes. Direct the oil to the far side of the piston. If compression improves, the problem is in the bore. Otherwise, it's probably the valves.

in firmly and give it a half turn. This helps make sure it seals. Turn the engine over the same number of strokes for each cylinder tested—about five will do—and observe the maximum gage pressure. Record the results and the procedure until you've tested all cylinders in order—just to keep things organized. Now that you have the numbers, what do they mean?

You may see pressures anywhere from 80 to 250 psi, depending on the engine and its problems. The best approach is to compare the cylinder pressures with each other. All cylinders of an engine don't go bad at once, so the weak ones will show up like a "sore thumb." Another thing to keep in mind: cylinder pressures of a high-compression engine—such as the Cobra Jet and Super Cobra Jet 429 engines—may read lower than their low-compression cousins at cranking speeds. This is due to cam timing and not because of their mechanical compression ratios. The same doesn't apply at higher RPM. Therefore, if you have a high-performance engine, don't be shocked if your neighbor boasts of higher cylinder pressures from his tamer regular-gas-burning engine.

What should the "spread" be between cylinder pressures? A good rule is the lowest pressure should not be less than 75 percent of the highest. Otherwise, something is not quite right in your engine. For example, if the highest reading is 150 psi and the lowest is 120 psi, then the engine is all right because 150 × 0.75 = 112—the lowest allowable pressure.

Now what do you do if all your readings are not within the 75-percent range? Something is wrong, but what? To test the piston rings, squirt about a teaspoon of heavy oil in the bad cylinder/s from your oil can—40 weight will do. The best way of determining if you are putting in the right amount of oil is to see how many squirts it takes to fill a teaspoon. Squirt the same amount of oil in the spark-plug hole while directing it toward the far side of the cylinder. It will also help if you have the piston part way down the cylinder to ensure getting the oil all the way around the piston. Don't get any oil on the valves. The oil will take a little time to run down around the rings, so wait a couple of minutes before rechecking pressure. If this causes the questionable cylinder/s to increase in pressure, the rings and bore are at fault and a rebuild is in order. If the pressure doesn't change an appreciable amount, look further.

Leak-Down Testing—Leak-down testing is similar to compression testing, but rather than the engine doing the compressing, a *leak-down tester* does it. Compressed air is used to apply a *test pressure* via a special fitting through the spark-plug hole to the cylinder being tested. The tester monitors the pressure the cylinder can maintain in relation to the test pressure. This is the best way to test an engine's condition. It eliminates factors which affect the results of a conventional compression test, but do not reflect the sealing qualities of a cylinder: valve timing, camshaft wear or engine cranking speed. The problem with a leak-down tester is it costs about three times as much as the conventional compression tester and requires a compressed-air source. Consequently it's not practical for you to purchase such an expensive piece of equipment for a "one-shot deal."

Leak-down testing is done by most tune-up shops. A leak-down tester is an integral part of Sun Electric's electronic analyzer. Sun recommends that an engine which leaks 20% or more of its test pressure needs attention, whereas a good cylinder will have 5 to 10% leakage. So, if leak-down testing your engine finds a problem cylinder, you'll know the problem will be rings, valves or a head gasket. It could also indicate a crack in the cylinder head or cylinder wall.

Head Gasket or Valves?—Two candidates

Leak-down testing is the most-accurate way of measuring a cylinder's sealing capability because it eliminates extraneous factors which affect compression-gage readings. Because leak-down testing is done without cranking the engine, cranking speed, valve timing and cam condition do not affect the reading. A leak-down tester requires a compressed-air source. Photo courtesy Sun Corporation.

may be causing the problem at this point—a leaky head gasket and/or a valve. If two cylinders are down on pressure and they are adjacent to one another, chances are good the problem is a blown head gasket between the cylinders. It could still be a head gasket even if only one cylinder is affected. A blown gasket is nearly always accompanied by an unusual amount of coolant loss from the radiator as cylinder pressure will leak into the cooling system. This situation is easy to diagnose. Remove your radiator cap and look at the coolant while your engine is running and warming up. If cylinder pressure is escaping into it, you'll see bubbles. Before making this check, make sure your coolant level is up to the mark. If you see bubbles, smell the coolant. You should be able to detect gasoline or exhaust fumes as the bubbles surface and burst if they are caused by a blown head gasket.

A sure way to test for this is to take your car to a professional who has a device which "sniffs" the coolant. It indicates whether or not the bubbles are cylinder gases and not just recirculating air bubbles. If you determine it is a bad gasket, you'll have to remove a head to replace it. Use the procedures outlined in the engine teardown and assembly chapters for this job. Check the head and block surfaces for flatness. Fix any problems or you may end up having to repeat the job.

Now for the valves. If you didn't find your compression-loss problem with the piston rings or a head gasket, the last probable culprit will be a valve/s. There are numerous reasons for valves leaking. A valve may not be fully closing or it may be burned. Both result in an unsealed combustion chamber. If a fully closed valve leaks, it's probably burned. So check for full closure first. Unless the problem is severe, you'll need vernier calipers or a dial indicator.

Pull off the valve covers and locate the cylinder you want to check. It should be on TDC (top dead center) just at the beginning of its power stroke—not between its exhaust and intake stroke. This ensures both valves *should be* fully closed. To do this, trace the spark-plug leads of the cylinders you are going to check in the distributor. Put a mark/s on the distributor body in line with the terminals on the distributor cap. Remove the cap.

When you crank your engine over and line up the distributor rotor with the mark on the distributor, you'll know the cylinder will be reasonably near TDC on its power stroke.

If both valves are fully closed, you should be able to rotate the pushrods with your fingers. It's unlikely the cam is holding a valve open unless the valves have been misadjusted. I know of no instances where valve adjustment has gotten tighter. If adjustment changes, it gets looser. If the pushrod rotates, the rocker arm has unloaded the valve, so it is free to close. If it won't rotate, first check to make sure that cylinder is at TDC, ready to start the power stroke, then back off on the adjustment using the procedure in the engine assembly chapter.

429CJ and SCJ VACUUM LEAK

Here's a problem that can frustrate the best mechanic, and the solution can't be found in any shop manual. The problem is with the 429CJ and SCJ end cylinders—numbers 1, 4, 5 and 8. They won't fire at idle with the valve covers removed, only when RPM is increased. Chances are you won't run into this problem if you have a CJ. However, because SCJ's valves require hot lashing, you obviously have to run your engine without the valve covers. It will idle rough with one off and won't run at idle with both off! The reason is the end, or corner cylinders, fire in sequence—8, 1, 5, 4—compounding the problem. Four fire, then four don't.

The cause of this problem is a vacuum leak. I'll bet you think it has something to do with the valve covers being off? You're right. The 429CJ/SCJ intake ports are so large that the end valve-cover bolt holes in the top row are not blind like those in the standard 429/460 head. They go straight through into the number-1, -4, -5, -8 intake ports. This creates a huge vacuum leak when one of these bolts is removed. The fuel charge is leaned so much at low RPM that it won't fire. Thread the bolts back in while the engine is running—valve cover on or not—and the cylinders start firing. So, before running your 429CJ or SCJ without its valve covers, thread the bolts back in the holes immediately above the end intake ports.

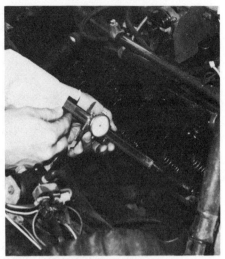

Checking valve lift by measuring how much a valve spring compresses from the full-open position to its closed position confirms whether or not your cam is in good shape. Refer to the chart when checking your valves.

One thing to be aware of is your engine probably does not have adjustable valves. Rather, it will use *positive-stop* rocker-arm studs or *pedestal pivots* unless your engine is a 351C Boss, 351C HO, 429 SCJ or 429 CJ built before 10-18-69. These engines had adjustable valves. Otherwise your valves must be *adjusted* using pushrods which vary ± 0.060 in. from the standard pushrod length. These special pushrods should not be required except during a rebuild when major machining is done such as valve face and seat grinding, or head and block resurfacing.

A word of caution, if your valve train is the same as what was installed at the factory it will have hydraulic lifters—unless you have a 351C Boss, HO or a 429 SCJ. Hydraulic lifters load the pushrods slightly, making them a little hard to turn. Therefore, don't let this fool you into thinking the valve in question is open. If your engine is equipped with a mechanical cam, the pushrods will be loose. That's why mechanical-cam engines are noisy. They are loose because the required clearance, or lash, is about 0.020 in. when hot—as opposed to the zero lash of hydraulic lifters.

CHECKING VALVE LIFT

If all the valves appear to be closing as just described, check the pushrods to see if they are too loose. If one is, it could mean a valve is hanging up in its guide,

Using a dial indicator to measure camshaft lobe lift directly is more accurate than checking at the valve.

preventing it from fully closing. To confirm this, you'll have to use your dial indicator to get an accurate reading on actual valve lift. The depth-gage end of a vernier caliper will work if you don't have a dial indicator, but it's not as easy to use. If you don't have either of these, use a 6-in. rule with a slide for measuring depth. It's not accurate, nor is it very easy to use, but you've got to use what you have.

If your engine has hydraulic lifters, make sure they are primed by cranking the engine. Then the valves will be getting as much lift as the cam can provide. Bump your engine over until the valve you're checking is fully opened, compressing its valve spring. If you are using a dial indicator, set the indicator plunger against the top of the spring retainer and in line with the valve stem so you'll get a true reading, and zero the indicator. If you're using a vernier caliper or scale, measure from the spring-pad surface—where the spring sits on the cylinder head—to the top of the spring retainer. Record the measurement. Bump the engine over until you can rotate the pushrod, indicating the valve is closed, and remeasure. You can read valve opening directly with the dial indicator. Repeat this a couple of times to make sure your figures are cor-

VALVE JOBS AND HIGH-MILEAGE ENGINES

Doing a valve job on a relatively high-mileage engine may not solve its oil-consumption problem as an engine's parts wear at the same time. As the sealing quality of the valves becomes less, the same thing is happening to the rings and pistons. After the cylinder heads are reconditioned, they will seal better, creating higher compression and vacuum loads on the rings and pistons. Where the rings and pistons were doing an adequate job before, they may not be able to seal satisfactorily after a valve job. Consequently, increased oil consumption and blowby may result. So beware of the valve-job-only solution.

CAMSHAFT LIFT (Inches)					
		Intake		Exhaust	
Engine	Year	@ Lifter	@ Valve	@ Lifter	@ Valve
351C-2V	70-74	0.235	0.407	0.235	0.407
351C-4V	70	0.247	0.427	0.247	0.427
351C-4V	71-72	0.247	0.427	0.247	0.427
351C-4V	73-74	0.277	0.480	0.283	0.490
*351C Boss	71	(0.290)	0.477	(0.290)	0.477
*351C HO	72	(0.298)	0.491	(0.298)	0.491
351C CJ	71	0.277	0.480	0.277	0.488
351M	75-79	0.235	0.407	0.235	0.407
351M Truck	77-79	0.250	0.433	0.250	0.433
400	71-79	0.247	0.427	0.250	0.433
400 Truck	77-79	0.250	0.433	0.250	0.433
429	68-69	0.278	0.486	0.278	0.486
429	70-73	0.253	0.443	0.278	0.486
460	68-69	0.278	0.486	0.278	0.486
460	70-79	0.253	0.443	0.278	0.486
460 Calif.	72	0.253	0.443	0.245	0.424
429CJ	70-71	0.289	0.506	0.289	0.506
*429SCJ	70-71	0.298	0.509	0.298	0.509

* Mechanical cam (solid lifters).
Lift shown in parentheses is approximate.

rect, particularly if there appears to be a problem with the valve sticking or a worn cam lobe. You should arrive at a valve-lift of 0.407–0.509 in. depending on which engine you have. Refer to the specification chart for the correct lift.

One problem with checking valve lift on high mileage hydraulic-lifter engines is the lifters are usually so worn that they can't maintain sufficient pressure to hold the valve completely open very long. As for how long, you'll know by the length of time your car sits before its valve/s clatter upon start-up. They leak down and partially close the valve with the result being a false valve-lift reading. So, be aware of this if your engine fits this category.

What if valve lift is not up to specification? You are checking for a sticking valve, so this is the first thing to suspect. As a double check, the installed height of the valve spring when the valve is closed should measure about 1.81 inch between the spring pad and the *underside* of the spring retainer. Check the valve-spring specification table on page 86 for your engine. If spring height is less than specified, the valve is not closing for some reason. As a final check, back off the rocker-arm-stud nut or pedestal bolt until the rocker arm is loose. It should be already if the valve is stuck. If spring height doesn't change, this confirms a sticking valve. You can remove the head with some assurance that a valve job is in order. If the spring or valve lift measures up to specification, recheck cylinder pressure, but only after adjusting the valve or installing a shorter pushrod and warming the engine. See pages 117-118 for valve adjustment. If it comes up to specification, you've found the problem.

Checking Camshaft-Lobe Wear—If you've found a valve is not lifting to specification, but it's properly adjusted and the installed spring height is right, suspect a bad lobe on your cam. A worn cam lobe usually has nothing to do with causing low cylinder pressure, but it has a lot to do with power loss.

If your engine is down on power, check the lift of all valves with a dial indicator. Rather than using the spring retainers to check from, check each cam lobe directly via its pushrod for the most accurate measurements. You'll have to loosen all the rocker arms and rotate them out of the way of the pushrods first. Starting with the number-1 cylinder, mount the dial indicator so it lines up with the pushrod. You may need a piece of tape around the dial-indicator plunger and pushrod to hold them together.

Make sure the pushrod is well seated in its lifter and the lifter is solidly against the base circle of the camshaft lobe. With the dial indicator set at zero, slowly bump the engine over with the starter while you watch the indicator. Note the maximum indicated reading and record it with the cylinder number and note if it's an intake or exhaust valve. You'll need this information later for comparison with the other lobes.

Camshaft lift at the lobe is the difference between the highest portion of the lobe and the diameter of the *base circle* as indicated by the sketch on page 63. Lifts vary from a high of 0.298 in. to a low of 0.235 in., depending on the engine you have, and the lobe you're checking. Lobes which operate the intake valves on the 429/460 engine have less lift than the exhaust lobes. Refer to the specification chart on previous page.

When checking camshaft-lobe wear, you won't have much trouble distinguishing bad ones from good ones. When a lobe begins to wear, it goes quickly. It doesn't wear gradually like a cylinder bore or valve guide. Differences won't be in thousandths, usually they'll be in tenths of an inch. All the lobes of a camshaft don't all wear at once. They go one at a time once the hardened lobe surface is worn through. It's not uncommon to see only one worn cam lobe while the rest are perfectly OK.

Regardless of how many lobes you find worn, if you find any, replace the cam and *all* the lifters. Otherwise, the new cam will be ruined before you get your vehicle out of the driveway. Use the procedure outlined in the engine assembly chapter when changing a camshaft and its lifters.

One final note before proceeding to the next chapter. If you've discovered a bad cam, but your engine's compression is good and it doesn't use much oil—no more than one quart per 1500 miles—install a new camshaft and lifters. However, if your engine is using up to 500 miles per quart and the compression is down below 75 percent of the highest on some cylinders, it is time to rebuild, including a new cam and lifters.

> **FIRING ORDER**
> Making assumptions can get you into trouble when it comes to diagnosing, tuning or building an engine. One of the most common errors is the assumption that all Ford V8s have the same firing order—not so! 351C, 351M and 400 engines have the same firing order: 1-3-7-2-6-5-4-8. 429 and 460 fire: 1-5-4-2-6-3-7-8. Firing orders are cast into the top of the intake manifold.

2 | Engine Removal

Because you are reading this chapter, I assume your engine needs rebuilding and, consequently must come out. Pulling an engine is one of the most troublesome and potentially dangerous parts of rebuilding an engine. These troubles are compounded when it comes time to reinstall the engine. So, a careful and orderly removal job avoids or minimizes problems at *both* ends of the project.

PREPARATION

Removing an engine is like diagnosing one—to do it right you'll need special equipment other than the standard set of tools. You should have something to drain engine liquids into, a jack and jack stands to raise and support your vehicle, a method of lifting your engine, a couple of fender protectors and some masking tape for identifying loose ends so you can tell where they go when it's time to reinstall your newly rebuilt engine. Remember some hand cleaner too.

Before starting the engine-removal process, ask yourself some questions. First, is your lifting device and what it will hang from strong enough to support 700 pounds? A chain hoist hanging from a two-by-four is not. Also, will you be able to leave your car where the engine is removed? Finally, will you be able to move the engine or car once the engine is lifted free of its compartment?

Lifting the Engine—One of the more common methods of pulling an engine is to do it in a garage with a chain hoist slung over a cross beam. The car is jacked up and supported by jack stands, or driven up some ramps, followed by getting the engine ready for removing. After the engine is hoisted out of the engine compartment, the car is set back down on the ground and rolled outside to an out-of-the-way place to wait for its rejuvenated engine. The engine can then be lowered to the floor for teardown. The order is reversed at installation time. If ramps are used, it is tough to push a car back up on them.

Unfortunately, a too-common result of this method is the money saved by doing the rebuild job yourself can be negated by the expense involved in rebuilding the garage roof—or worse yet, paying the hospital bills. If the chain-

351 Cleveland is one of the best Ford engines ever. It fills the bill as an excellent passenger-car engine and as one of the best high-performance engines ever to come out of Detroit.

hoist support is strong enough, the drawback is you'll have to move your car before lowering the engine—and once the engine is lowered you'll have to move it. A final word of caution about using this method: I've never ever found a conventional passenger-car garage with a beam sturdy enough to support an engine safely, so make sure your health *and* life insurance premiums are up to date if you're going to try it anyway.

Then there is the "shade-tree" approach. Set up an A-frame made from 3 12—15-foot-long, 5- or 6-inch diameter poles—preferably under a tree for shade, of course. Chain them together securely at the top and hang a chain hoist from the chain. A *come-along* will also do the job. Drive your car up two ramps located so the engine ends up directly under the chain fall or come-along. Block the wheels so the car won't roll back down the ramps about the time you start pulling your engine. Get the engine ready for removal and lift it up and clear of the car, then roll the car off the ramps and lower the engine to the ground. Even though I'm jesting about the shade-tree method, an A-frame like I've just described is a lot stronger than a garage beam.

The most convenient device for removing an engine is a "cherry picker." One of these can be rented from your nearest "A-to-Z" rental on a daily basis—about $10 a day. Most of them can be towed behind a car. You'll need one for an hour for removing, and again when installing your engine. This neat device lets you lift your engine out, then move it where you're going to do your teardown.

With my sermon over, let's get on with pulling your engine. Because the 351C, 351M/400 and 429/460 engines have been installed in so many types and descriptions of vehicles, in addition to the complexities involved in explaining how to pull an engine out of all these vehicles, I'll generalize. Let your common sense fill the voids. The vehicle I used as an example is a 1974 Ford equipped with an accessory-loaded 400. The only thing it didn't have was an air-injection pump.

It's easier to work with a clean engine. A can of spray degreaser and your garden hose or the local car wash will do the job. Cover the carburetor to keep water out. A wet ignition will probably be your biggest problem if you have to restart your engine.

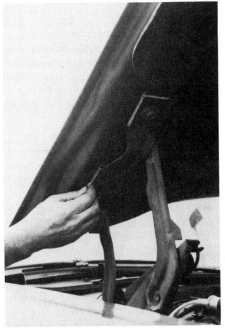 

To ensure that your hood will fit at installation time, trace around the hinges for later reference. I'm using machinist's *soap-stone* here. With the front hinge bolts removed, finish removing the rear bolts and lift the hood off. Don't forget to disconnect the engine-compartment light if your vehicle is so equipped.

Before immobilizing your vehicle, clean its engine compartment, engine and transmission. Remove as much dirt and grease as possible. The most effective and simplest way of doing this is with a can of spray degreaser and high-pressure water. If your car is running, take it to a car wash and just follow the directions on the can to soak the grease and dirt before spraying it. Or, use their car wash degreaser if one is available. Otherwise, do it at home with a garden hose. Take fair warning: what's on your engine ends up under the car afterwards, so act accordingly. After you've finished your engine, it may not look new, but it will be a whole lot easier to work on.

With your engine and the surrounding area clean and everything ready to go, I'll explain how to remove the engine point-by-point.

Fender Protector—Put a fender protector, or a suitable facsimile over both fenders to protect the finish. This also makes the fender more comfortable for leaning on, and your tools won't slide off as easily.

Battery—If you have a standard transmission, remove the battery and store it in a safe place. Otherwise, disconnect the ground cable at the battery, but leave the battery in place because it will come in handy later on. Disconnect the coil-to-distributor lead—the one coming out of the center of the distributor cap, or the distributor-to-amplifier lead.

Remove the Hood—If you have an engine compartment light, disconnect it first. Before loosening the bolts, mark the hinge locations relative to the hood. Do this by tracing around the hinges on the hood with a grease pencil, some chalk or a scriber. This will save you the trouble of readjusting the hood after you replace it.

A neat method of guaranteeing a hood goes back in exactly the same position that is was before its hinge bolts are loosened is to drill an 1/8-in. hole up through each hinge and the hood inner panel—*don't go through the hood!* To reinstall the hood, bolt it loosely to the hinges, insert an ice pick or an awl through the holes to align it, then tighten the bolts. The result—perfect alignment.

Remove the front hinge bolts and loosen the back ones while supporting the hood at the front. A helping hand comes in handy—you on one side and him on the other. Remove the back bolts and lift the hood off. Place it out of the way where it won't get damaged. Stand it up against a wall, and to ensure it won't fall over, wire the latch to a nail driven in a wall stud. Another trick is to put it on the roof of your car. Protect the paint or vinyl top by putting something between the hood and the roof. Use this method only if your car or truck will be parked inside out of the wind.

Remove the Radiator First—Turn your attention to the fan and radiator. Removing the radiator provides more access to the front of your engine for removing the accessories. This also prevents the radiator from being damaged during engine removal. One little nudge from the engine as it is being pulled can junk a radiator.

The first impulse is to remove the fan before the radiator, but this is a sure way to remove skin from your knuckles. The core fins put a radiator in the same family with a cheese grater.

> **PLAN AHEAD**
> It's very tempting and easy to "go crazy" disconnecting everything in sight without regard to where things have to go later on. *Don't rely on your memory*. Even if there are only 5 hoses there are 120 ways they could be installed, and 119 are wrong! Mark *all* the items you disconnect such as wires and vacuum hoses. Wrap masking tape around the end of the wires or hoses and make a little "flag" for writing where the ends go. Put similar flags on the connection itself so you can get it all back together. Put your camera to use. Periodically snap some pictures of your engine from a couple of angles as you disconnect things. A picture *is* worth a thousand words. Finally, use tin cans or small boxes for the bolts, nuts and washers. If your engine is loaded with accessories, you'll find it especially helpful to label the containers as to the particular unit the fasteners were used with.

Avoid an underfoot mess by draining your radiator before disconnecting any hoses. Loosening the radiator cap speeds up this process.

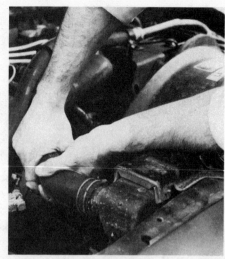
Remove top and bottom radiator hoses. After loosening the clamp, a little twist will break the bond so the hose can be pulled off.

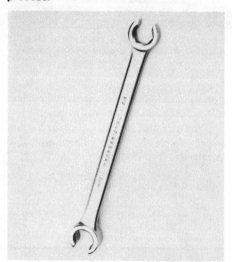

Use a flare-nut wrench when disconnecting automatic-transmission-cooling or fuel lines to prevent rounding the nuts. I'm using an another wrench to hold the union which threads into the radiator. After disconnecting the cooling lines, connect them with a hose to keep transmission fluid from siphoning out of the transmission.

With fan shroud loose and laid back over the engine, free the radiator. With the upper bracket/s removed, radiator can be lifted out. However, if you have a setup like this down-flow truck radiator, securing bolts through its side flanges have to be removed.

Start by draining the radiator—it's faster with the cap off. Remove the top and bottom hoses and replace them if they are more than a couple of years old, particularly if you live in a hot-dry climate like the southwest U.S. Disconnect the automatic-transmission cooling lines which run to the bottom of down-flow radiators and to the side of cross-flow radiators. Use a tube-nut wrench to prevent rounding off the nuts. This wrench looks like a wide six-point box-end wrench with one flat cut out so you can slip the wrench over the tubing, then onto the nut. After loosening the nuts, slide them back from the ends of the tubes and connect the two lines with a hose to prevent transmission fluid from siphoning out and messing up your driveway or garage floor. A clean hose the size of your radiator over-flow hose will do nicely, in fact you can use it if it's clean.

The fan shroud is next if you have one. Unbolt if from the radiator and lay it back over the fan and on the engine. To protect the delicate radiator fins *and* your knuckles, cut a piece of cardboard to fit the backside of the radiator core, then tape it in place. If your radiator is the down-flow type which is bolted solidly to the radiator support, there are 4 bolts, 2 at each side. As you remove these bolts, be ready to support the radiator and lift it out. Do it with care because of the easily damaged fins. If you have the rubber-mounted cross-flow type radiator, remove the bracket/s which clamps over the top, and lift the radiator out. While you still have hold of it, store the radiator in a safe place where it can stay until you're ready to reinstall it. The trunk is a good place. Better yet, have your radiator reconditioned at a radiator shop, then store it. Remove the fan shroud.

Fan—With the fan out you have a much clearer view of the front of your engine. Loosen the fan-attaching bolts and lift fan, spacer and bolts out together. On clutch-drive fans, you'll need an open-end wrench to get to the bolts between the fan and pulley.

If your fan is of the viscous-drive or clutch variety, store it face down. Otherwise fluid will leak from the clutch if it is stored with its mounting flange on the down side. If enough fluid leaks out you'll end up with a free-wheeling fan and a high replacement cost.

Air Cleaner—Remove the air cleaner after disconnecting the hoses and ducts which attach to it such as the hot- and fresh-air ducts, and the crankcase-vent hose.

Radiators can be heavy like this big-capacity truck unit. Lift yours out carefully so as not to damage the fins—they bend easily.

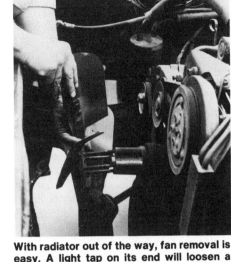
With radiator out of the way, fan removal is easy. A light tap on its end will loosen a stuck fan spacer.

Disconnecting the carburetor linkage. Cable-type throttle linkages pop right off by using a screwdriver like I'm doing here. Put any clips you removed back on carburetor to avoid losing them.

Look at all that "spaghetti." Masking-tape "flags" with hose or wire locations written on them make engine installation a whole lot easier.

If your power-steering cooler is mounted to the top of the A/C compressor, you'll have to work it out from under the A/C lines before removing the pump. Power-steering pump, lines and cooler can be removed from the engine and laid over to the side out of the way. Make sure pump sits right-side up so fluid doesn't drain out.

Throttle Linkage—It will either be the rod-and-lever type or the cable type. If your engine is equipped with the rod-and-lever linkage, disconnect it from the carburetor, then wire it out of the way to the dash panel (firewall). On cable types, disconnect the cable at the carburetor also and pinch the tabs or remove the retaining screw at the manifold bracket so the cable can be withdrawn. If your engine is backed up with an automatic transmission, also disconnect the TV rod, or transmission kick-down rod from the carburetor. Wire it up out of the way also. Replace the clips on the carburetor so they don't get lost. They are special, consequently you could have a difficult time finding replacements.

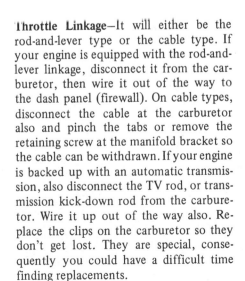

When removing last A/C-compressor mounting bolt, be ready to handle 25 pounds. Lay compressor aside, or tie it up out of the way.

ENGINE REMOVAL 15

Don't Trust Your Memory—Use a camera and/or some masking tape. Label each hose and wire *before* disconnecting it. Disconnect all the hoses and wires from the top of the engine. There'll be a hose from the power-brake booster to the intake manifold, or one to a vacuum manifold mounted on the firewall and many smaller hoses, depending on the year and how your vehicle is equipped. An engine-wiring harness usually lays along the inboard flange of the left valve cover, held by clips under two or three valve-cover bolts. Disconnect the harness leads from the oil-pressure sending unit, water-temperature sending unit, coil and/or distributor, A/C compressor clutch and any emission-related devices. Remove the harness with the valve-cover clips, or if they are the type which bend over for retention, bend them open to release the harness. Disconnect the heater hoses. Plan on replacing them if they are more than two-years old.

ACCESSORY REMOVAL

Pulleys and Belts—Engine accessories can be the most difficult part of removing and replacing an engine. Loosen all the belts and remove them. This will free the water-pump pulley, but you may not be able to remove it if your engine has a three-belt accessory-drive system. You'll have to remove the crank pulley first. After the crank-pulley bolts are out, tap on the backside of the pulley using a rubber or plastic mallet to loosen it. The pulley pilots over the center of the crankshaft damper, so it may be a little tight. The water-pump pulley will now be free to come off.

A/C Compressor and Power-Steering Pump—I'll describe an engine with a complete array of accessories, so disregard those areas which don't apply to your engine. Start with the A/C compressor and power-steering pump. Remove the triangular bracket which sometimes is installed between the compressor and the power-steering brackets. With this bracket out of the way you'll be able to see most of the power-steering bracket bolts. Before removing the A/C compressor, you'll have to remove the power-steering pump and its bracket because it mounts to the bottom compressor bracket. Unbolt the pump bracket from the block, being careful to support the pump right-side up to prevent fluid loss. Leave the hoses connected to the pump and set the pump aside. Wire it to the left-front fender apron if you can't position the pump so it won't fall over and spill its fluid.

Avoid Disconnecting the A/C Hoses—Here's where you can save time, trouble and money. Rather than disconnecting the A/C-compressor hoses for removing the compressor from the engine compartment, set it aside like the power-steering pump. This avoids the need to recharge the A/C system after you've reinstalled your engine.

Begin removing the compressor from the engine by first removing the upper support bracket—the one that also mounts the adjustable idler pulley. Follow this up by unbolting the compressor and lower bracket from the front of the engine as an assembly, then lay it to the side out of the way. If your engine compartment is small, such as the Mustang's or Cougar's, you'll need to get the compressor and its lines up out of the way. Support the compressor from the left fender with a strong cord tied to a bent nail. Tie the free end of the cord to a bolt threaded into the compressor. Lift the compressor and bracket high on the fender apron, then hook the nail to the wheel-opening flange with a rag or a piece of cardboard under the nail. This will keep the paint from being damaged. Tie the cord short enough to support the compressor high and out of your way.

Alternator and Air Pump—The left-front side of your engine now should be bare, so shift your attention to the opposite side. There will be an alternator, and possibly, an air pump—used with Ford's *Thermactor* emission-control system. One will be mounted above the other, but which way depends on the year and model. If yours has an air pump, remove it after disconnecting the hose. Also remove the bypass valve and other related hardware. As for the alternator, don't remove it yet as it'll have to be in place for a little while longer. You can remove its adjusting bracket and loosen its long pivot bolt.

A/C-compressor/power-steering-pump setup on a 460-powered Lincoln Mark IV. Rotary compressor is mounted topside on a common bracket with power-steering pump. Regardless of pump type and location, basics are the same.

Don't Forget the Filler Tube and Ground Straps—Two easy things to overlook until your engine ceases to move as you're pulling it out are the automatic-transmission filler tube and the engine ground straps. The filler tube and a ground strap (usually only one) are normally attached to the rear of the right cylinder head by a common bolt. If two bolts are used, remove the one retaining the filler tube. As for the ground strap, it'll be easier to disconnect it from the firewall rather than the engine. Put it back on the firewall after you have the engine out just so it doesn't get forgotten or lost between now and when you replace your engine.

Clutch Linkage—With a standard transmission using a rod-and-lever clutch linkage, you have some more up-top work. Some models use a return spring which attaches between the top of the equalizer bar and the firewall—disconnect it. Parallel to the spring is a pushrod extending through the firewall to the top of the equalizer bar. Disconnect it at the equalizer, being careful not to lose the bushing and retaining clip. If it looks worn out, count on replacing it. However, to keep

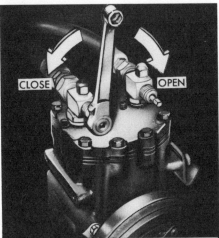
If you must disconnect A/C lines to remove the compressor and have a system equipped with service valves, close the valves as shown. This isolates the system so you can disconnect service valves with lines from the compressor side of valves. Photo courtesy Ford.

I'm removing the crankshaft accessory-drive pulley because it's one more thing to interfere with removing the engine, and it has to come off eventually.

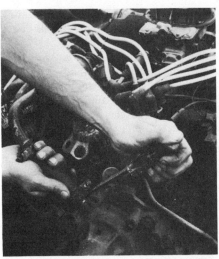
With the alternator and starter-solenoid leads removed, only thing left to come off the front of this engine from above are the heater hoses. Because metal tubes are used, hoses at water pump and block are so short they can't be twisted to break them loose. That's why I'm using a screwdriver.

SAVING YOUR AIR-CONDITIONING REFRIGERANT

If you must remove the A/C compressor prior to removing your engine, the hoses must be disconnected. This can be accomplished without losing all the refrigerant on earlier models with *service valves*. Service valves were phased out during the mid-70's as a cost savings. These valves are in-line with the hoses or lines, and are usually covered with cadmium-plated caps. Remove the caps and close the valves. Rotate the high- and low-pressure service valves at the compressor in the counterclockwise direction to isolate the charge from the compressor. *Before* removing the hoses with service valves intact from the compressor, loosen the gage-port cap a *small* amount until you are certain *all* the pressure has escaped. **CAUTION: High pressure in these lines can hurt or blind you if you are careless. Wear goggles to protect your eyes, even when bleeding pressure from the compressor.** Now you can remove the hoses by disconnecting the service valves from the compressor so they stay with the hoses and seal the system. *Don't remove the valves from the hoses.* Place the hoses out of the way and you're ready to remove the compressor after sealing it so it will stay clean and dry. Use the caps which are used to seal new compressors before they are installed. You should be able to pick some up free from your local garage.

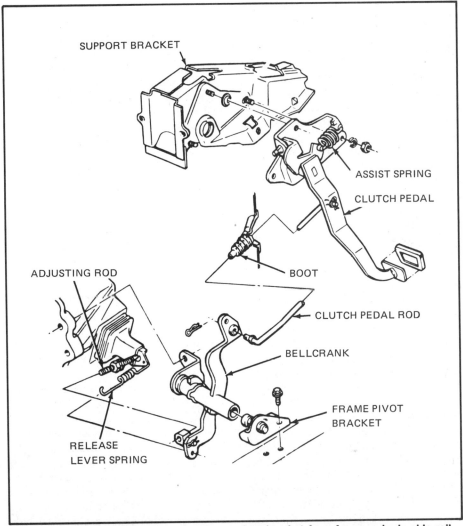
To remove clutch-linkage bellcrank, remove pivot bracket from frame or body side rail, then disconnect clutch-pedal rod and release-lever pushrod and spring from bellcrank. Bellcrank will then slide off pivot at the engine. Drawing courtesy Ford.

from losing it, put it back on the pushrod along with the clip. There may also be a *wave washer.* Put it back too.

The equalizer bar fits between the engine and the frame or body. It pivots on a bracket attached to the frame or body with 2 bolts. After removing these bolts and the bracket, the equalizer bar will be free to pull off its pivot at the engine, but you'll have to finish this job from underneath. This can usually be accomplished without raising your car by simply reaching the bellcrank-to-release lever under the car. Unhook it and the release-lever pushrod will swing free. You can now remove the bellcrank and pushrod together by sliding the bellcrank off its engine pivot.

Disconnect the Fuel Pump—*Before breaking any fuel-line connections,* make sure there are no open flames close by such as a gas water-heater pilot light. Disconnect the fuel-tank-to-fuel-pump line at the fuel pump. To prevent siphoning and the consequent fire hazard created by spilled gasoline, push a 3/8-in. diameter bolt into the end of the hose. Make sure the bolt isn't fully threaded. Otherwise fuel will leak out around the threads. It should have at least 1/2 in. of unthreaded shank.

Get What You Can From the Top—If the exhaust-manifold-to-exhaust-pipe bolts are accessible from above, particularly the top ones, get them now. They are definitely easier to remove this way unless you have a 12-inch extension in your socket set. Some penetrating oil on the threads about an hour prior to removing the nuts and the use of a box-end wrench help avoid damaged knuckles. An open-end wrench slips off too easily and there may not be enough room for a ratchet handle.

Engine Mounts—Ford did us a big favor by designing the engine mounts for easy removal and installation. Each mount is a two-piece assembly. One half stays with the engine and the other with the frame or body. The two halves are held together with a *through-bolt* and nut. Depending on your vehicle, you'll have to judge now whether it will be easier to remove the bolts now, or from underneath after you've raised your car. One hitch is—if you have a 351M/400 or 460 in a truck, you'll have to remove the *insulator* or rubber portion of the engine mount. You'll also have to remove the vertical bolts on the frame-side of the insulators. This has to be done because the engine can't be lifted high enough to clear the engine mounts and disengage from the transmission due to interference at the firewall. This is the most difficult por-

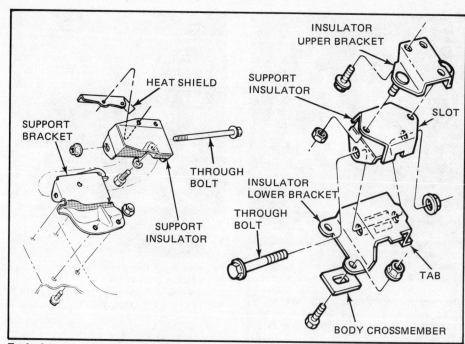

Typical two- and three-piece engine mounts. Both use single through-bolt. If yours is equipped with a sheet-metal heat shield, make sure you don't lose it. Direct exhaust-manifold heat will quickly destroy the rubber. Drawing courtesy Ford.

tion of removing and replacing these engines in trucks. The engine will have to be raised as high as possible, plus a little more, to get the insulators out from between the engine-mount brackets.

To undo the bolt and nut, place a box-end wrench on the nut and use a socket and ratchet with an extension on the bolt. As you start to unthread the bolt from the nut, the box-end wrench will rotate and stop against the engine or body or frame so you won't have to hold it. When the nut is off, slide the bolt forward out of its mount. I find it easier to get to the left mount from below. You'll have to assess your particular situation.

Jack Your Car Up—It's time to get your car in the air. A truck usually has enough ground clearance to work under without having to raise it. Not so with a car. A hydraulic floor jack and two jack stands are great to have at this point. You'll only have to raise the front of your car for removing the engine. To do this, place the jack under the number-two crossmember. Frame cars have a substantial crossmember to which the front-suspension lower control arms attach. Unit-body cars have a less-substantial-looking tubular crossmember which bolts to the body side rails. It is all right to use this crossmember to raise the car, but be careful because it's easy to get the jack pad under the steering linkage, possibly bending the center link. A 2"x4" wood block between the jack and crossmember helps.

With the car in the air, place the jack stands under the frame or body of the car rather than under the front suspension. I prefer this method because it's more positive. As you can see from the photo, I placed jack stands under the front frame *torque boxes.* For unit body cars, the sway-bar brackets are a good place. Whatever you use to support your car, make sure it's substantial and that you block the rear tires to keep your car from rolling.

Don't use bricks, cement blocks or cinder blocks as stands or supports. These materials work fine with evenly distributed loads, but they will crack and crumble when subjected to high-point loading as when supporting a car. If I seem to be dwelling on this subject too much, it's because when a car falls, the result is often fatal. *Be careful!*

Now's the time to get your trusty creeper into service—or a big sheet of cardboard works well, particularly if you're not on hard pavement. A creeper doesn't roll in dirt or gravel.

Finish the Clutch Linkage—With the car in the air, complete those partially finished jobs you started from the top. If you haven't already done so, the equalizer bar can be removed by disconnecting the retaining spring which fits between the release lever and the equalizer bar. This will free the equalizer bar so you can slip it off its engine pivot.

Exhaust System—Finish disconnecting the exhaust system from the manifolds. Here's where the long extension for your socket

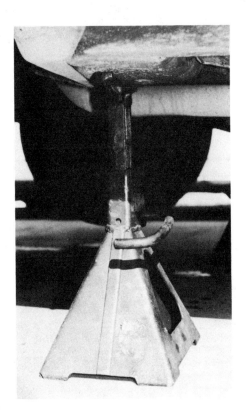

When jacking up your car, jack under number-two crossmember—under the engine. Some unit-body cars are not equipped with this crossmember, so you'll have to jack under something solid like the front-supension strut-rod bushing bracket. This car is firmly supported by jack stands under front torque boxes. Don't forget to block rear wheels.

set comes in handy for reaching the nuts. Don't be surprised if the studs unthread from the manifold.

Cooling Lines and Starter-Motor Cable—With an automatic transmission, make sure the cooling lines are unclipped from the engine. Some have clips and some don't. If yours does, just push the lines out of the clips to release them. While you're looking up where these lines go—usually on the right side with down-flow radiators and left side with cross-flows—make sure the starter-to-battery lead is out of its bracket.

Engine Mounts—If you couldn't reach an engine mount from topside to remove its bolts, do it now. Remember, once the nut is off, just slide the bolt out.

Expose the Converter—With an automatic transmission, remove the converter cover at the front of the engine plate. It's usually attached with two or three bolts.

Get Power to the Starter—With an automatic transmission, reconnect the battery ground so you'll have power to the starter. Now, if you don't possess a remote starter, have a trustworthy friend *bump* the engine over to expose each of the converter-to-flexplate nuts for removal—a socket on the front of the crank pulley works well too. Particularly if you've removed the spark plugs.

CAUTION—If you use the friend-and-starter method, make sure he understands the ignition switch *is not to be touched* without your direction.

Now for the Starter Motor—After removing the converter nuts, disconnect the battery cables at the battery. Remove the bottom starter bolt first, then the top one while supporting the front of the starter with one hand. Lifting a starter out is tricky on unit-body cars equipped with conventional steering—not rack and pinion. Slide the starter forward, and drop its nose—geared end—between the steering linkage and the converter housing. If you don't need to remove your starter motor for rebuilding, you don't need to disconnect the cable. Just leave it hanging there until it is ready to be replaced. Support it with some wire or rope to take the load off the cable.

Finish Unbolting the Bellhousing—Regardless of which transmission you have, the rest of the removal process is pretty much the same. The only job left before lowering your car is to remove the remaining bellhousing/converter-housing bolts. After doing this, take one last look underneath just to make certain *everything* is disconnected from the engine.

Advantage of a long extension for disconnecting exhaust pipes is evident here. A shot of penetrating oil on nuts and studs an hour or so before you're ready to remove them will also help.

ENGINE REMOVAL 19

Remove the Alternator—After setting your car down, the last thing to do before pulling the engine is to disconnect the battery ground and alternator leads from the engine if you haven't already done so. Remove the alternator and set it aside out of the way. You may have to support it with some wire or rope. Remove the battery.

Lift the Engine Out—Take a last look around the engine compartment to double-check that everything is disconnected—inevitably, there will be something. Once you're confident the engine is ready, position your lifting device over the engine, or the car and the engine under it. Attach a lifting cable with looped ends or a chain to the front of one head and the rear of the other with bolts and washers. Or if you're lucky, your engine will be equipped with lifting lugs the factory used for installing your engine.

They will be attached to the cylinder heads with the exhaust manifolds and their bolts.

Make sure there is enough slack in the cable/chain so you'll be able to set the engine all the way down once it's out of the engine compartment. The chain/cable must be short enough so the engine can be lifted high enough to clear the body work when pulling it out, so keep this in mind also. Hook onto the chain/cable approximately in the middle so the engine will be balanced once it is free from the transmission and engine mounts. Lift the engine high enough so the mounts will clear as the engine is moved forward. Bring the jack in contact with the transmission. Now you're ready to disengage the engine from its transmission. This may require a little juggling. Standard transmissions require moving the engine farther forward so the input shaft fully disengages from the clutch. Regardless of which transmission your car is equipped with, the engine will first have to be lifted high enough so it can be moved forward. Keep the jack in contact with the transmission as the engine is raised.

Don't force the engine if it hangs up. You could damage something. Check around to see what bolt, ground strap or other connection you may have overlooked. Correct the problem and proceed. As soon as the engine is completely free of the transmission, hoist it out.

Mark the Converter and Flex Plate—With light-colored paint—spray or brush—mark the stud at the bottom of the converter and the matching bottom hole on the engine flex plate. Flex-plate holes and

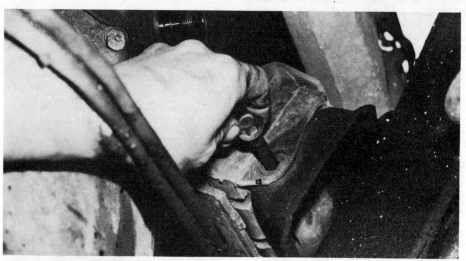
With nut off, engine-mount bolt is free to slide out.

With front-cover plate removed, converter-to-flexplate mounting studs and nuts are exposed. Index engine with starter or a wrench on damper bolt to bring each stud and nut into view. Don't forget to mark one stud and flexplate so they can be reinstalled in their original positions.

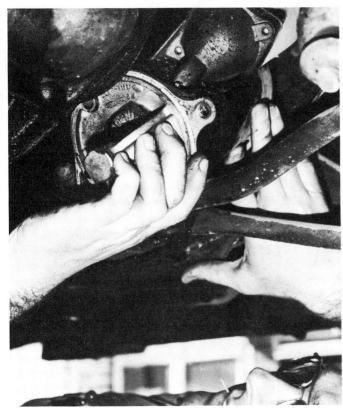
After starter-motor bolts are out, slide the starter forward out of housing, then lower its nose to remove it.

Lifting engine out. As you can see, hydraulic cylinder is fully extended so engine won't go up any higher. We cleared the sheet-metal by standing on front bumper to lower car, then steered engine past hood ornament.

torque-converter studs have a very close fit, so mark them so they can be installed in the original position when the time comes. This is one variable you'll know is right if things don't click together at installation time.

Secure All the Loose Parts—If you don't want to leave your jack under the transmission or want to free your car so it can be rolled, support the transmission so it doesn't fall when you lower the jack. Wire coat hangers work well for this job and for supporting the exhaust system.

With your engine out and hanging in mid-air, begin stripping it down as described in the teardown chapter. You did remember to drain the oil, didn't you? Before the parts are scattered, collect and store them in a secure place where they won't get mixed in with your lawn mower, motorcycle, snowmobile or any other parts you may have lying around. You'll be glad you did. If you get most of the grease, oil and dirt off the parts and put them in containers, the trunk of your car or bed of your truck is a good place to store them.

Continue with your engine teardown by skipping the next chapter and go directly to Chapter 4. Save the Parts Identification chapter for when you're ready to sit down and relax.

Transmission and exhaust system is supported by wires hanging from a steel angle bridging the hood hinges. This eliminated need for jack to support the transmission and allowed moving the car.

ENGINE REMOVAL 21

3 Parts Identification and Interchange

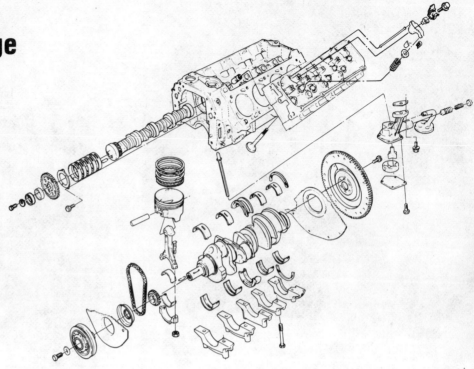

351C major components. What you know about interchanging parts from another engine to yours can save you considerable money. What you don't know can cost you dearly. Drawing courtesy Ford.

It's a real money-saver to know which parts can be interchanged, especially when you need to replace one of your engine's major components with parts from another engine.

Will a 351C crank work in a 351M, or will 429 heads fit on a 460? How do you identify the parts if they will? Armed with this information you can visit the local junkyard knowing your options rather than being forced to purchase new parts for your specific engine—or having to play Russian Roulette with parts.

You need to know exactly which engine you have before you order parts or try to decide what parts will interchange with it. You need to know the *displacement*, *year* and *change level*. Change level is Ford's way of identifying a change made in the midst of a model year.

IDENTIFICATION

Which Engine Do You Have?—To start from ground zero, I'll assume you don't know which engine your car is equipped with. Look at the identification tag under the left-front corner of the windshield. The five-digit alphanumeric group contains model year and engine-code information. The first number is the year—77, 78, etc. You have to know the decade. The last letter is for the engine. You must use both the year *and* the engine code to determine which engine was installed in the car. For example, all 351-2V engines have the same code H. In this instance this gets even more complicated. U.S. production of the 351C ceased in '74 and the 351M began in '75, so there's no overlap. However, the 351W overlaps them both. Consequently, you'll not only have to use engine code and model year to identify your engine in this case, you must recognize its physical features if the engine-identification sticker is not on the air cleaner, or there is no air cleaner.

The major external appearance difference between the 351W and 351C/351M engines is the cylinder head. 351W heads are smaller. Because you won't have one to compare against the other, their relative sizes can't be used for identification. Cylinder-head exhaust-manifold mounting surfaces are angled differently: 351W exhaust-manifold surface is almost vertical; 351C/351M surface is angled 45° to vertical, or generally in line with the side of the cylinder block. This makes the 351C/351M head wider between the intake-manifold and exhaust-manifold mounting surfaces. Those two surfaces are parallel on the 351C/351M heads.

Engine Tag or Decal—Once you've determined which engine you have, you'll also need to know its date of production, change level and engine-code *number*—this is different from the engine's code letter on the vehicle identification tag. The engine-code number used by the Ford parts man when he's ordering parts is found on a tag or decal attached to the engine. Up through February 1973, tags were mounted at the front of the engine. They are usually sandwiched under the coil-mounting bracket, dipstick-tube bracket or under the water-temperature sending unit. Unfortunately, when a tag is removed for one reason or another, they are usually discarded rather than being replaced—a *big mistake*. Beginning in February 1973, the metal tags were replaced in favor of a decal on the right-valve cover at the front. Use the sketches to figure out how to read a tag or decal.

If your engine tag or decal is missing, that's a problem! But, there is a way out. Somewhere on your engine is stamped an *alphanumeric group* which is the engine *build date* code. It can look something like 9A12C. The first digit is the calender year, 1969 or 1979. The next letter is the month starting with A for January and continuing through M. I is not used. The last two digits are the day, the 12th of the month in this case.

You can find the build-date code on the machined surface on the front of your engine block, called the *front-face*. On the 351C and 351M/400 you'll find it stamped at the lower-right corner; at the upper-right on 429/460s.

Casting Numbers—When a component such as a cylinder head or block is cast, a number is cast in the part and is appropriately called a *casting number*. Casting numbers are extremely helpful when identifying an engine or its parts. Unfortunately, casting numbers are not 100-percent accurate because castings are frequently machined differently, thus generating different *part numbers* which don't appear on the part! To make matters worse, it is conceivable these parts won't interchange. Consequently, a part's casting number in conjunction with its physical design must be used to identify it. So let's get on with the physical makeup of the various engines and their parts.

You'll also find numbers similar to the build-date code cast into your block and heads once you get your engine apart. These indicate the date on which the parts were cast. Note that the casting date is always prior to the build date.

DESIGN DIFFERENCES

Before getting into each engine's parts in detail, let's take a look at the basic differences between the 335 and 385 Series engines, and a couple of differences within the 335 Series.

335s Versus 385s—The major point is that *none* of the major 335 and 385 engine components interchange. Even though some of their components are similar in design, they are entirely different engines. The 429/460, or 385 Series, is much larger. This is illustrated by a comparison between their *bore spacings,* or distance between the center of one bore and adjacent bore/s. The 335 Series has a 4.38-inch bore spacing, whereas the 385 is approximately 1/2 inch more at 4.90 inches. This is a major size and weight contributor as the block, cylinder heads, crankshaft and camshaft all must be longer.

In the "up-and-down" department, the *deck heights*—center of the crankshaft main-bearing bores to the block deck surface—of the 351M/400 and 429/460 engines are similar at approximately 10.3 inches. The 351C measures 9.21 inches—a difference of more than 1.00 inch.

The major *design* difference between the two engine series is how coolant is routed from the block and cylinder heads to the radiator. The 385 Series is basically the same as the small-block Ford. Coolant is routed from the block, through the heads and into the intake manifold, making it a *wet manifold*. The thermostat and its bypass are in the manifold. Coolant travels from the intake manifold, through the thermostat, then to the radiator when the thermostat is open. When it is closed, coolant recirculates through the engine, around the thermostat via the thermo-bypass and to the water pump, consequently not passing through the radiator.

The 335 Series does it differently. 351C and 351M/400 intake manifolds are *dry*. No coolant flows through them. Coolant is routed internally through the cylinder block through the top of the integral front timing-chain cover to the thermostat and radiator, or the water pump. The thermostat and water-pump bypass are also in the block. Coolant is routed through the thermostat to the radiator, or through the thermostat bypass via the *bypass orifice*

ENGINE CODES

Code	Engine	Comments
H	351C-2V	Same as 351M-2V Code
Q	351C-CJ	
M	351-4V	
R	351C Boss	
R	351C HO	
H	351M-2V	Same as 351C-2V Code
S	400-2V	
K	429-2V	
N	429-4V	
P	429 Police	
C, J	429CJ	'J' indicates Ram Air
CV, CY	429SCJ	'CV' indicates drag option
A	460	
C	460 Police	73-78

If your engine is original equipment, you can identify it by the code on the vehicle-identification tag on the left-front door-latch face or hinge pillar, or the serial-number plate on top of the instrument panel at the far left side. Cross-reference this code to the above chart. Drawing courtesy Ford.

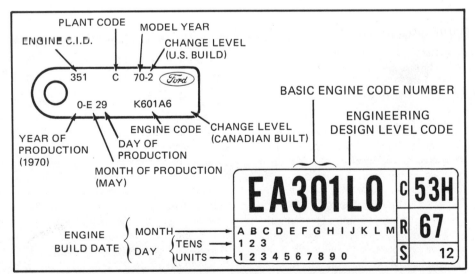

Engine-identification tags were used up to February 1, 1970, then were replaced by decals like that shown at right. You'll need engine displacement, model year, change level, date of production and engine code number when ordering engine parts.

PARTS 23

UNDERSTANDING THE PART NUMBERS

If you are like most people the first time you cast your eyes on a Ford part number, you probably thought it looked unnecessarily complicated. However, this first impression is gradually changed as you understand what all the numbers and letters stand for.

Engineering and Service Numbers—At the risk of confusing you, you should first know that there are two numbers for each *finished part*: the *engineering*, or *production* number and the *service* number. They are just what they say. The engineering number is assigned by engineering when a part is approved for production. This number is used by the assembly plants. The service number is assigned when the part goes into the part-distribution system. It is a different number because how a part is finished and packaged for service is different from its original production counterpart. It's the number used by your Ford parts man to look for or order. He doesn't want to know about the engineering part number. However, you may need to because the engineering number is the one appearing on many parts.

Casting Numbers—Casting numbers are special engineering numbers placed on a casting to assist in identification by the plant. They are cast on a part at the time of manufacturing, and apply only to the basic casting. One casting can be machined to make slightly different parts, thus creating several different finished engineering- and service-art numbers. Using casting numbers to identify a part is sort of like playing horseshoes. It doesn't count much when you're close, but it does count—and it very well may be the only number you have to work with.

Dissecting the Numbers—Ford part numbers, regardless of whether they are engineering or service numbers, always consist of three distinct groups: *prefix, basic part number and suffix.*

Engineering Casting Number	Production Number	Service Number
D0VE-6090-C	D0VE-6049-C	D0VZ-6049-D

Prefix—The four-digit *alphanumeric* prefix tells the year the part was released by engineering for production, the car line the part was originally released for and by what Ford engineering division (chassis, engine, body, etc.). In the case of a service part, the prefix tells which Ford car division

VEHICLE IDENTIFICATION CODES

- A — Full-size Ford
- B — Bronco 70–73
  - Maverick 75–77
- C — Torino Elite
- D — Falcon 60–69
  - Maverick 70–74
  - Granada 75 and up
- E — Pinto
- G — Comet 61–68
  - Montego 69–76
- I — Monarch
  - Versailles
- J — Industrial engines
- K — Edsel
  - Comet 60–61 and 75 and up
- L — Lincoln 58–60
  - Mark III, IV, V
- M — Mercury
- O — Fairlane 62–68
  - Torino 69–76
  - LTD II 77 and up
- Q — Fairmont
- S — Thunderbird
- T — Truck
- U — Zephyr
- V — Lincoln 61 and up
- Z — Mustang

the part is for—Ford or Lincoln–Mercury. Example: the preceding part numbers are for a 429 cylinder head originally released in 1970 as indicated by the first letter—D for the 70s, C for the 60s and so on. The following number is the year in that decade—0. The third digit, usually a letter, indicates the car line the part was originally designed for. The accompanying list gives most of them.

The V designation is given to 429/460 engine parts because the engine was designed originally for Lincoln cars. The last digit, or letter in the prefix in this part number indicates the part was released for production by the Engine Division: A is for chassis, B is for body and E is for engine—makes sense because it's an engine part. This applies to engineering parts, however the digit in a service part refers to the car division—Z for Ford Division, Y for Lincoln-Mercury Division and other letters for special parts such as X for the extinct Muscle Parts Program.

Basic Part Number—Regardless of whether it's an engineering or a service part number the basic part number will be the same. For example, 6049 is for *all* cylinder heads, 6303 is for crankshafts and 6010 is for blocks. The number for the casting used for machining these parts is different. Referring to the cylinder head again, the basic *finished* part number is 6049 whereas its casting number is 6090. Because it is relatively easy to put a casting number on a part while it is being cast, it's the casting number that appears on the part—great for them and terrible for the guy trying to identify a part. Also, the number that appears on a casting may not include the basic casting number for the simple reason that you don't need a number to tell you you are looking at a block or an intake manifold. The number generally consists only of the prefix and the suffix, or D0VE-C for the cylinder head.

Suffix—A part-number suffix generally tells you the *change level* of a part, regardless of whether it is applied to the casting, the finished part or the service part. A signifies a part produced as it was originally designed, B indicates it was changed once, C twice and right through the alphabet in sequence, excluding the letter I. When the alphabet has been gone through once, the suffix grows to two letters and starts as AA, AB, AC and so on. How does a change affect the other two numbers? A service part and its number can change independently of the casting part and its number and the engineering part and its number simply because it comes after these two in the scheme of things. Using the same reasoning, a finished, or engineering part can change independently of the casting, but not of the service part. A casting affects both the finished and the service parts. This is why the suffixes of all three numbers rarely ever match.

24 PARTS

Cylinder-block, crankshaft and cylinder-head casting numbers. Block casting numbers are found under rear of right cylinder bank. Casting date on this 351C block is there also. 429 and 460 cylinder-head casting numbers are on ledge above exhaust ports. 351M/400 numbers are in the rocker area under the valve cover and 351C casting numbers are under the intake ports. Crankshaft casting numbers are on side of front throw as on this 429 crank.

into the backside of the water pump. The 351C uses a removable bronze disc as a *metering orifice:* it is cast into the 351M/400 engine. This orifice *meters,* or restricts the flow of coolant bypassing the thermostat *and* radiator during times that the thermostat is open. This ensures that enough coolant circulates through the radiator for cooling. The same thing is accomplished in the 429/460 by the diameter of the bypass and its connector hose.

Balance—Another significant design difference between the two engine series is, 335 engines are *externally balanced;* 385s are *internally balanced.* Internal balance means that nothing needs to be added to the crankshaft *externally* to balance the engine's rotating and reciprocating components—piston, connecting-rod and crankshaft assemblies. 335 engines require additional *counterbalancing* at their crankshaft dampers and flywheels. Therefore, you can't install a 429 flywheel on a 400 and expect it to be balanced, even though the flywheel will fit both engines. The engine will not be *in balance.* Vice-versa, a 400 flywheel will put a 429 *out of balance.*

Let's look at the individual pieces in detail and pay particular attention to their similarities and differences.

BLOCKS

351C—All 351C cylinder blocks are virtually the same with one notable exception. The '71 Boss, '72 HO, '71 CJ and '72–'73 4V versions use four-bolt main-bearing caps. 351C-4Vs produced through 1971 and all 351C-2Vs use two-bolt mains. However, these blocks are cast to accept four-bolt main caps. If you have four-bolt caps, a two-bolt block can be easily converted to a four-bolt block as the cap registers are the same. What's required is some precision machining. The outboard bolt holes must be drilled and tapped, then the main-bearing bores are line bored and honed with the caps in place. Although four-bolt blocks are stronger,

Engine build date code 1K25—Oct. 25, 1979—stamped on front face of 351C block indicates engine was assembled five days after the block was cast. See block casting date 1K20 under casting number in top left photo.

Unique 351C bellhousing bolt pattern. 400 at right shares same bolt pattern with the 351M and 429/460 except for 1973 400 installed in full-size Fords with FMX automatic transmission (transmission code PHB-AD). This 400 and all 351C patterns are the same as six-bolt small-block Ford. 1.09-inch higher deck height of 400 is also evident.

PARTS 25

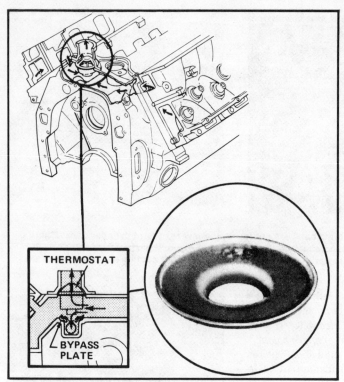

Integral front timing-chain cover cast into 351C block houses thermostat and water-pump bypass orifice which are in 429/460 manifold. 351M/400 is identical to the 351C with one exception: A removable orifice plate is used for the 351C, whereas the 351M/400 is cast integrally with the block. If this plate is left out for one reason or another, too much coolant will bypass the radiator, resulting in overheating. Drawing and photo courtesy Ford.

Eccentric weight on this crankshaft damper is part of "external" balancing for the 351C and 351M/400. Rest of external balancing is done at flywheel. 429/460s are internally balanced, consequently you won't see eccentric weights on their dampers or flywheels.

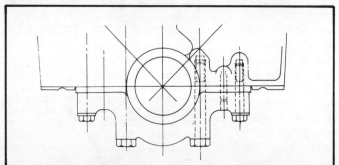

Drawing shows four-bolt main-bearing caps like those on the 351C Boss, HO, '71 CJ, '72 and later 4Vs and the 429CJ and SCJ engines. Photo shows 351C-2V block which is cast to accept four-bolt caps (arrow). Extra wide main-bearing-cap register will accept a four-bolt cap. Two-bolt 429/460 block has to be used with two-bolt caps. Four-bolt 429/460 block is a unique casting. Drawing courtesy Ford.

there is complete interchangeability between all 351C blocks. Block casting numbers are under the rear of the right cylinder bank.

351M/400—I'm discussing the 351M and 400 blocks at the same time because they share the same block. The difference lies in their "lower-end" components—connecting rods, pistons and crankshafts. This is not the case when comparing the 351C to the 351M/400. Even though they are in the same engine family, their blocks are not interchangeable. To begin with, the 351M/400 deck height is 1.09 inches higher than the 351C block—9.206 inches versus 10.297 inches. Although both 351s have the same bore/stroke combination, their main-journal diameters are different at 2.75 inches for the 351C and 3.00 inches for the 351M/400. In addition, the 351M/400 connecting rod is 0.80 inch longer than the 351C, 6.58 inches compared to 5.78 inches measured center-to-center.

Distinguishing the 351M/400 block from the 351C is relatively easy due to their deck-height difference. The 351C's intake-manifold mounting surface is nearly flush with the top of the integral timing-chain cover. However, it's approximately 3/4 inch higher on the 351M/400 block. In addition, the 351M/400 has a 1-inch-high web extending straight out from the front of the manifold surface, just left of the distributor hole.

Another significant difference between the 351C and 351M/400 blocks is their bellhousing/converter-housing bolt patterns. Both have a 6-bolt pattern, but they are arranged differently with one exception. Consequently, if you attempt to install a 351M/400 in the place of a 351C, you won't be able to without a transmission change too. I know, I've been bitten by this one. The exception is

When sifting through your local scrap yard for parts, make sure you inspect carefully before laying down your money. Two 429 engines are good "for instances." Very desirable four-bolt block is complete with a broken connecting rod. However, if a block is all that's needed, this one would be a good deal as the block appears to be OK. As for the other 429, its block is scrap as shown by crack. The coolant wasn't drained. All it took was one sub-freezing night.

the 400 installed in a full-size '73 Ford with the FMX automatic transmission (code PHB-AD). It has the same pattern as the 351C!

A final note about the 351M/400 block. It was cast at one of two Ford foundries, the Cleveland Foundry or the Michigan Casting Center. Those cast at the Michigan Casting Center prior to March 2, 1977 experienced water-jacket cracking problems immediately above the lifter bores due to an internal coring problem. The cracks will be horizontal and approximately one inch above the top of the lifter bores. To determine if a block may have the problem, look for the casting-date code. The magic date is March 1, 1977, or in code form 7C01. This has to be cross-referenced with the casting-plant code—CF for Cleveland Foundry and MCC for Michigan Casting Center. These codes are located on the top-rear of the block next to the oil-pressure sending-unit hole. So, if the block in question has a 7C01 or earlier date code and the MCC plant code—an M above two Cs, one inside the other—look closely for cracks.

429/460—Although totally different engines, the 429/460 has the same bellhousing pattern as the 351M/400, but not the same as the 351C. All of the blocks are the same with exception of the main-bearing caps. Some are two-bolt and some are four-bolt caps. All 429 and 460 blocks use two-bolt main-bearing caps except for '70–'71 SCJs and '71 CJs whose blocks are cast *and* machined to accept four-bolt caps in the number-2, -3 and -4 positions. Because the two-bolt 429 and 460 blocks are not cast to accept four-bolt caps, they can't be machined to accommodate them.

With the exception of the main-bearing caps, the only other interchangeability consideration is *deck height,* the distance from the centerline of the crankshaft bearing bores to the block deck surface. 429/460 blocks originally had a 10.300-inch deck height. It was increased 0.010 inch to 10.310 inches in '70-1/2 for a slight reduction in compression. A significant reduction in compression ratio was made in 1972 by a combination of increasing deck height to 10.322 inches and increasing cylinder-head combustion-chamber volume. This resulted in a 2-point drop in compression—from 10.5 to 8.5.

Four-bolt main bearings and deck-height differences don't present *mechanical* interchangeability problems, however there are other considerations. It would be a step down to go from a four-bolt block to a two-bolt one if strength is what you're after. Because of deck heights and cylinder-head combustion-chamber volumes, block and cylinder-head combinations produce different compression ratios. You should be aware what compression ratio you'll end up when switching heads and blocks. This is covered in more detail later in the chapter.

You'll find 429/460 cylinder-block casting numbers under the right-rear cylinder bank. The number is vertical, or parallel to the engine block's rear face.

CRANKSHAFTS

335 and 385 Series crankshafts have one thing in common: all are cast-iron. Otherwise, their dimensional differences prevent crankshaft interchanges between 351C, 351M/400 and 429/460 engines.

351C—351C main- and connecting-rod-journal diameters are 2.749 inches and 2.311 inches, respectively. As with all Ford 351 engines, the stroke is 3.50 inches and the engine is externally balanced. There are differences between 351C cranks. Boss and HO crankshafts are selected for their *high-nodularity* content. *Nodular* cast-iron has its graphite in the form of spheres rather than the flakes of *grey* cast-iron. High nodularity means the cast-iron contains more nodules, or spheres, resulting in higher strength. Boss and HO cranks can be identified by a *Brinell* test mark, a spherical indentation approximately 1/8-inch wide in the front or rear face of one crankshaft counterweight. The rough casting surface is ground smooth in the area tested, so it shouldn't be too difficult to find the test mark if it is there. Remember, the test only ensures the strength of the crankshaft. Consequently, most other 351C cranks will have the same strength, the test just proves it for high-performance applications. Regardless of this one difference, 351 crankshafts are interchangeable and can be identified by code number 4M or 4MA on the side of the front counterweight.

351M/400—351M and 400 crankshafts are the same in all respects with one exception—*throw*. Crankshaft throw is the center distance between the main and connecting-rod journals. It is *half* the stroke. 351M and 400 strokes are 3.50 inches and 4.00 inches, respectively. Because stroke is the only exception, 351M and 400 crankshafts are interchangeable *if* 351M pistons are used with 351M cranks and 400 pistons are used with 400 cranks. Journal diameters are 3.000 inches and 2.311 inches, respectively.

429/460—Except for the numbers, the 429/460 crankshaft description sounds like the 351C. CJ and SCJ versions of the 429 use hardness-tested cranks which can be identified using the same method as for the Boss 351C and HO cranks: a Brinell test mark on the smoothed face of a counterweight. Main-journal and connecting-rod-journal diameters are 3.00 inches and 2.50 inches, respectively for both 429 and 460 cranks, however the 460's stroke is 0.30 inch more at 3.85 inches. Because their rods are the same, you can interchange crankshafts so long as the pistons are designed to go with the crank. 429 and 460 code numbers are on the side of the front counterweight.

CONNECTING RODS

All Ford 335 and 385 Series engines use forged-steel connecting rods with 3/8-inch connecting-rod bolts.

CYLINDER-HEAD CASTING NUMBERS AND SPECIFICATIONS (BASIC NUMBER 6060)

Engine	Year	Casting Number	Combustion-Chamber Volume (cc)	Valve Diameter (Inches) Intake/Exhaust	Comments
351C-2V	70-74	D0AE-E, J D0AZ-A, B, D D1ZE-CB D1AE	76.2	2.04/1.66	
351C-4V	70-71	D0AE-H, R	62.8	2.19/1.71	Valves and porting larger than 2V head for increased flow.
351C-CJ -4V	71 72-74	D1ZE-DA D1ZE-GA	75.4	2.19/1.71	Open-type combustion chamber and induction-hardened exhaust-valve seats, otherwise same as early 4V.
351C Boss HO	71 72	D1ZE-B D2ZE-A	66.1 75.4	2.19/1.71 2.19/1.71	Equipped with adjustable rocker arms and pushrod guide plates. HO uses open-type combustion chamber, otherwise same as '71 Boss.
351M/400	75-80	D5AE-AA D5AZ	78.4	2.04/1.66	
400	71-74	D1AE-A D3AE-G2B	78.4 78.4	2.04/1.66	
429/460	68-71	C8SZ-B C8VE-E D0VE-C	75.8	2.08/1.66	Positive-stop rocker-arm studs and rail-type rocker arms.
	72-74	D2VE-AA D3VE-AA, A2A	91.4	2.08/1.66	Non-adjustable stamped-steel rocker arms.
	75-78	D4VE	96.2	2.08/1.66	
429CJ, SCJ and Police	70-71	D0AE-H C9VE-A	73.5	2.24/1.725	SCJ and early CJ use adjustable stamped-steel rocker arms. Pushrod guide plates retained by screw-in studs.
460 Police	73-74	D34E	89.0	2.19/1.66	Same rocker-arm setup as 429 Police.
	75-78	D4VE	96.2	2.08-1.66	Same as passenger-car casting.

Casting numbers listed above and on preceding page are correct according to the best available information. Use physical dimensions to make final parts identification.

Infamous 351M/400 water-jacket-cracking problem in real life (arrow). Blocks cast before March 2, 1977 at the Michigan Casting Center (see the casting-plant code) are the ones with the problem. This cylinder block meets all the criteria.

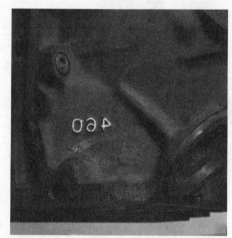

Just because it says 460 doesn't mean it is, mirror image or not. This is a 429. All it takes to make it a 460 is a 460 crankshaft and its pistons.

CYLINDER-BLOCK CASTING NUMBERS & SPECIFICATIONS (BASIC NUMBER 6015)

Engine	Year	Casting Number	Bore (Inches)	Deck Height (Inches)	Comments
351C-2V -4V	70-74 71	D0AZ-D D0AE-J, G D2AE-CA	4.00	9.206	Two-bolt main-bearing caps. Can be machined to accept 4-bolt caps.
351C Boss, '71 CJ, HO '72-74 4V	72-74	D2AE-CA	4.00	9.206	Four-bolt main-bearing caps. Originally equipped with split-lip rear main-bearing seal. Special block cast with thicker cylinder walls and siamesed bores for racing use carries an XE192540 casting number.
351M	75-80	D5AZ D7TE-A2B D8	4.00	10.297	
400	71-80	D1AE-A, AC, A2C D4AE-B2A D5AZ D7TE-A2B D8	4.00	10.297	
	'73	D3AE-B	4.00	10.297	Bellhousing bolt pattern same as 351C.
429/460 429CJ	68- 70	C8VE-F C8VY-A C9VY-A C9VE-B D0SZ-A, D D1VZ D1VE D1ZE-AZ D5TE	4.36	10.300; '68-'70 10.310; '70-1/2 10.322; '72-	
429SCJ CJ Police	70-71 71 71-72	D0OE-B	4.36	10.300	Four-bolt main-bearing caps in number 2, 3 and 4 positions.

CRANKSHAFT CASTING NUMBERS AND SPECIFICATIONS (BASIC NUMBER 6303)

Engine	Year	Casting Number	Stroke (Inches)	JOURNAL DIAMETER (INCHES)		Comments
				Main	Connecting Rod	
351C	70-74	4M 4MA	3.50	2.749	2.311	Boss and HO hardness tested. Brinell test mark on polished face of counterweight.
351M	75-80	1K	3.50	3.000	2.311	
400	71-80	5M 5MA 5MAB	4.00	3.000	2.311	
429	68-78	4U 4UA	3.59	3.00	2.500	
460	68-78	2Y 2YA 2YAB 2YABC	3.85	3.000	2.500	
	79-	3Y	3.85	3.000	2.500	

PARTS 29

429 four-bolt '70—'71 SCJ and '70 CJ engine block. End caps are still of the two-bolt variety, whereas center three use four bolts. Photo courtesy Ford.

Compression height of two pistons is different, so same connecting rods can be used in the same block with different stroke crankshafts. Compression height of 460 piston is 0.130 inch shorter than 429 piston.

351C—All 351C connecting rods are the same except for the Boss and HO bolts and nuts and the method used to inspect these rods. Boss and HO rods are fitted with high-strength (180,000 psi yield strength) bolts and nuts, and are final inspected by Magnafluxing® to ensure they are free of cracks. Otherwise, a 351C connecting rod is forged steel, it has a center-to-center length of 5.78 inches, a big- and small-end diameter of 2.4365 inches and 0.911 inch, respectively and *spot-faced* rod-bolt-head seating surfaces. The bolt-head seating surface is machined so it describes a radius on the inboard-side of the bolt head as viewed from the top of the bolt head similar to the 429CJ/SCJ pictured below. This reduces stress concentration on the rod at the bolt heads.

351M/400—The 351M/400 engines use the same connecting rods regardless of the engine or year. They have a 6.580-inch center-to-center length, small-end diameter of 0.9738 inch, big-end diameter of 2.4356 inches and the rod is spot-faced at the bolt-head seating surfaces.

429/460—Although there is more than one 429/460 connecting rod, their vital statistics are identical. They are forged-steel, have a 6.605-inch center-to-center length, 1.309-inch small-end diameter and 2.6529-inch diameter at the big end. What distinguishes the different 429/460 connecting-rod assemblies is the machining of the basic connecting-rod forging and the rod-bolt design. Although there are two connecting-rod forgings, C8VE-A and D0OE-A, the type of machining and the rod-bolt design are the major things to be aware of.

All early 429/460 connecting-rod bolt-head seats were broached. Rectangular-head bolts were installed in these assemblies. To increase the fatigue strength of the rectangular bolt head, the bolt-head-to-shank radius was increased in 1972. This change didn't make the bolt or rod suitable for heavy-duty or high-performance use. Consequently, a new connecting rod with spot-faced bolt-head seats and bolts with "football" heads were used. These rods were installed in 429 CJ and SCJ, 460 truck and 429 and 460 Police engines. Use of the broached rod was continued in standard 429/460 engines.

With all this connecting-rod information in mind, consider three choices. If your engine is for everyday passenger-car use, the broached rod with rectangular-head bolts are adequate. However, if you are preparing your engine for a heavy-duty or high-performance application, use spot-faced rods with football-head bolts. The third alternative is in the event you have the broached rods, but would like to upgrade them. Replace the rectangular-head bolts with those having the football heads. They won't be as strong as the spot-faced connecting-rod assemblies, however they will be stronger than the standard units.

PISTONS

As with most 335 and 385 engine components, the pistons are similar, but different in one or more respects—piston diameter, wrist-pin diameter or compression height. They appear similar because most have two intake-valve reliefs. However, the differences become apparent when it comes to interchangeability.

351C-2V, -4V, and CJ—These cast-aluminum pistons are the same, consequently they are fully interchangeable between 2V, 4V and CJ engines. Their flat tops are only disturbed by two intake-valve reliefs.

351C Boss and HO—Boss and HO versions of the 351C pistons are quite different from the standard piston design. They are

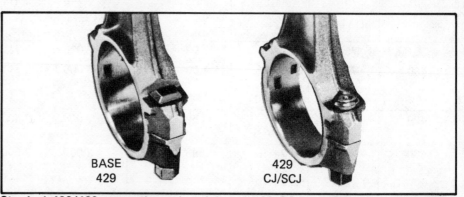

Standard 429/460 connecting rod on left and 429 SCJ and CJ rod on right. Only difference between them is machining at their bolt-head surfaces—broached versus spot-faced.

Pop-up '71 351C Boss piston and flat-top '72 HO piston give compression ratios of 11.3:1 and 9.2:1, respectively. Both are forged, whereas standard pistons are cast. Piston at right is TRW's forged version of the standard 351C cast piston.

forged for added strength. Additionally, the Boss uses a *pop-up* type 11.3:1 piston. A pop-up, or *dome* is a *lump* formed into the piston top to reduce combustion-chamber volume for increased compression. The '72 HO, a low-compression version of the '71 Boss, uses a flat-top forged-aluminum 9.2:1 piston. It is similiar in appearance to the cast version of the 351C piston.

351M/400—351M and 400 pistons are cast-aluminum, fit the same diameter bores and are flat-top with two valve reliefs. The major difference is *compression height,* the distance from the wrist-pin-bore center to the top of the piston. Compression height for the 351M is 1.947 inches. The 400 is shorter at 1.647 inches to compensate for its longer stroke. As a result, these pistons are not interchangeable. Remember, the 351M and 400 engines use the same connecting rods. Consequently, a 351M piston installed in a 400 would stick out the top of its bore or crunch against the bottom of the cylinder head as the piston approaches TDC. On the other hand, a 400 piston installed in a 351M would stop far short of the top of its bore at TDC and would hit the crankshaft at the bottom of the stroke.

429-2V, -4V and 460-4V—The 429/460 piston description is similar to that of the 351M/400. They are cast-aluminum, fit the same bores, have two intake-valve reliefs and have different compression heights. 429 and 460 engines also use the same connecting rod. The 460's pistons are dished to achieve the desired compression ratio. I'll discuss how an engine's compression ratio is determined later.

429 CJ, SCJ, Police and 460 Police—Other than the 460 Police, 429, CJ, SCJ and Police pistons differ from the standard piston. The 460 Police engines use conventional passenger-car pistons, 429 CJ and Police-car pistons are cast as *originally installed,* however their design differs from that of the standard 429 piston. Rather than having two intake-valve reliefs, 429 CJ, SCJ and Police-car pistons have one valve relief. Consequently, an originally installed piston must be installed in the same cylinder bank it came in because it has an offset wrist pin—the pin centerline is 0.0625 inch to the right of the piston center-line. This means the piston can't be turned around to clear the intake valve for installing in the opposite bank as is possible with a piston having no pin offset. Fortunately, all 429 CJ, SCJ and Police *service* pistons are manufactured without this offset making it possible to install the piston in either direction. Ford services the CJ and 429 Police piston with the SCJ piston. Its design is the same with one major exception—it's forged.

CRANKSHAFT, ROD AND PISTON MATCHING

What if you're replacing your connecting rods, pistons and crankshaft with those from another engine? Some interchanging is possible due to the 351M/400s and the 429/460s using the same blocks, however this is about it. To do any interchanging, you must be aware of the basics. Otherwise, you can end up with a basket of expensive junk.

For instance, crankshaft throw, connecting-rod center-to-center length, piston compression height and deck clearance must be compatible with the block's deck height. For a given stroke crankshaft, a longer rod will cause a given piston to operate higher in its bore. A shorter rod would cause the same piston to operate lower in the same bore. Piston compression height is very important here because of the 351M/400 and 429/460 situations. As compression height is increased—the pin bore is moved down in relation to the piston top—the top of the piston will operate higher in the cylinder bore. Reducing compression height moves it down. The last item, deck clearance, is simply the measurement from the piston top at TDC to the *block deck* or cylinder-head gasket surface. This clearance is necessary to prevent a piston from hitting the cylinder head. The question is, "How does all this relate to swapping parts?"

The *sum* of the crankshaft throw, connecting-rod length and piston compression height *must not exceed a block's deck height,* otherwise serious engine damage *will* result. This is particularly true if it is exceeded by more than the specified deck clearance. In this case the piston will come higher than the deck surface and impact against the cylinder head. On the other hand, if the sum is considerably less, you'll have a very underpowered engine due to a too-low compression ratio.

One final and vital factor that must be considered when changing bottom-end components is *balance.* You don't want undue engine vibrations. Chances are this won't be a problem as the components that can be interchanged are balanced to work with each other, however it certainly wouldn't hurt to have bottom-end components checked for proper balance. This is particularly important in high-performance applications. Many engine machine shops are equipped to do this, and at very reasonable prices.

VALVE-TRAIN DESIGN

Variations in 335 and 385 valve trains are restricted mostly to the rocker-arm designs and how they are stabilized, or *guided*. To operate a valve, a rocker arm must *rock* only in the plane of the valve stem and the pushrod. Ford uses two basic rocker-arm styles and guides them in three different ways.

Valve-Stem-Guided—1968–71 429/460s are the only 335/385 engines using valve-stem-guided rocker arms. The rocker-arm pivot or fulcrum is spherical at its rocker-arm contact surface. Consequently, the only thing keeping the rocker arm from pivoting in all directions is the rocker-arm *rails,* thus the name *rail-type* rocker

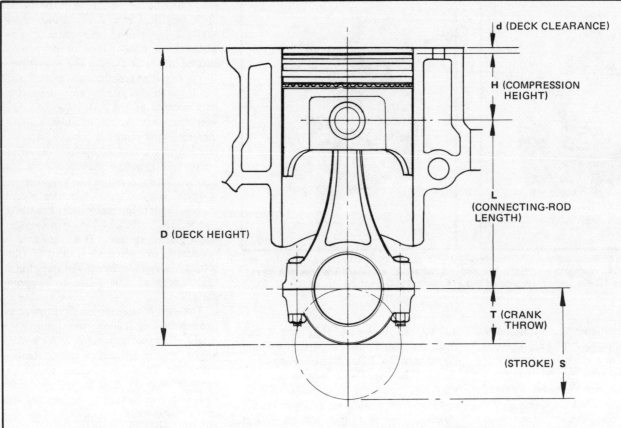

	351C 70-74	351C Boss 71	351C H.O. 72	351M 75-80	400 71-80	68-70½	429 70½	72-80	429CJ & SCJ 70-71	68-70½	460 70½	72-78
D	9.206	9.206	9.206	10.297	10.297	10.300	10.310	10.322	10.300	10.300	10.310	10.322
d	0.035	0.045	0.019	0.016	0.565	0.010	0.020	0.032	0.010	0.010	0.020	0.032
H	1.650	1.631	1.657	1.947	1.650	1.890	1.890	1.890	1.890	1.760	1.760	1.760
L	5.780	5.780	5.780	6.580	6.580	6.605	6.605	6.605	6.605	6.605	6.605	6.605
T	1.750	1.750	1.750	1.750	2.000	1.795	1.795	1.795	1.795	1.925	1.925	1.925

Crankshaft throw, connecting-rod length, and piston compression height must add up to no more than a block's deck height, otherwise serious engine damage will result. When interchanging parts, use accompanying chart to make sure your engine will have sufficient deck clearance.

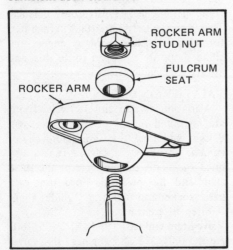

1968—71 429/460 rail rocker arms are stabilized by rails which straddle their valve tips. These are the only rocker arms in 335 and 385 engine series which are cast-iron and use spherical fulcrums. Note also, the positive-stop stud. Drawing courtesy Ford.

arm. Flanges cast into the end of the rocker arm straddle the valve-stem tip. This restricts lateral rocker-arm movement. The problem with this set-up is additional loads are put on the valve-stem tip, the valve guide and the rocker arm. As a result, all wear a bit more than normal. Rather than the guide wearing toward and away from the fulcrum due to the normal sliding motion of the rocker-arm tip against the valve-stem tip, it also wears sideways, or 90° to normal guide wear.

As rocker arm and valve-stem tip wear, the rails get closer to the valve-spring retainer. This doesn't present any particular problem until the rails begin to bear on the retainer. When this occurs, the retainer keepers are unloaded and can be released, letting the valve *drop* into its cylinder. Severe engine damage usually results.

Be wary of using worn valves and rail-type rocker arms in your engine. If the rails are close to the retainer—0.060 inch or less—don't take chances. Replace the rocker arm *and* its valve.

Pivot-Guided—Beginning with the 351C in 1970, all except the high-performance versions of the 335 and 385 engines use pivot-guided, stamped-steel rocker arms. Rather than using a ball, or spherical fulcrum, a cylindrical-shape pivot is used. This restricts the rocker arm so it pivots around one axis or in a single plane—the plane of the valve stem and pushrod. To do this, the pivot must also be restricted from moving. This is accomplished by registering the fulcrum in a notch in the cylinder-head rocker-arm pedestal and securing it with a bolt through the center of the pivot. Consequently, the rocker arm cannot be adjusted, but this is not

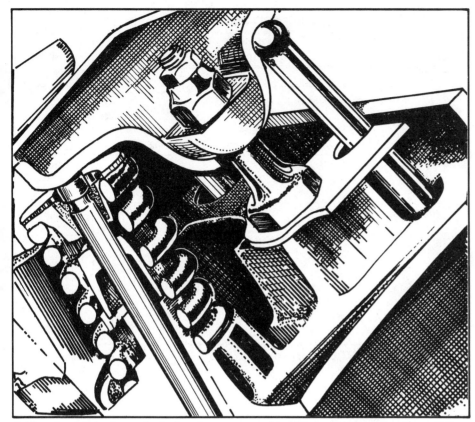

Push-rod-guided, stamped-steel rocker arms used in high-performance versions of the 351C, 429 and 460 engines. Drawing by Tom Jakeway.

required with hydraulic lifters. Pivot-guided rocker arms are superior to rail-type rocker arms because there is no stabilizing load on the valve. Therefore valve-stem, valve-stem-tip, guide and rocker-arm wear are reduced.

Pushrod-Guided—Pushrod-guided rocker arms are the high-performance versions of the 335 and 385 stamped-steel rocker arms. Their use is restricted to the 351C Boss, and HO, 429 CJ, SCJ and Police and early 460 Police engines. The difference lies in how the rocker arm is guided and how the pivot is fitted to the head. The cylindrical fulcrum is fitted to a rocker-arm stud.

The rocker arm and its pivot must not rotate around the stud. This is where the pushrod-guided aspect comes in. This is accomplished by pushrod *guide plates*. One guide plate controls two rocker arms and pushrods, so there are four guide plates per head. Each guide plate is secured under two rocker-arm studs. The pushrods operate in slots at the edge of their guide plates, thus preventing lateral pushrod and rocker-arm movement.

Engines with this type of valve train *must use hardened pushrods*. Conventional pushrods would quickly gall and fail.

Adjustable and Non-Adjustable—There are two classifications of rocker arms, adjustable and non-adjustable. Valve-stem and pivot-guided rocker arms are non-adjustable. Depending on the specific engine, pushrod-guided rockers arms may or may not be adjustable. 351C Boss, HO and 429 SCJ engines have adjustable rocker arms to allow setting valve lash with a mechanical camshaft and its lifters. Even though 429 CJ's built prior to 11-1-69 had hydraulic lifters, they used adjustable rocker arms, but changed to the non-adjustable *positive-stop* stud as used on '71–'72 429 and '73–'74 460 Police engines, and the early rail-rocker-arm equipped 429/460 engines.

The difference between adjustable and positive-stop rocker-arm studs is the adjustable stud has a 7/16-20 thread on a 7/16-inch stud, whereas the positive-stop has a 3/8-inch thread on the same size stud. This results in a shoulder at the bottom of the thread which the rocker-arm nut bottoms against, *fixing* the position of the stud and its *adjustment*. Without this shoulder and a longer thread, the adjustable rocker-arm nut can be moved up and down for adjusting its

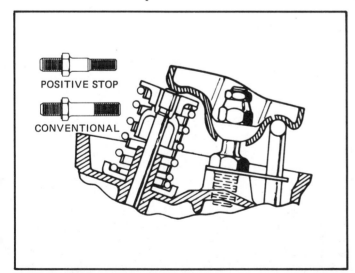

Section through pushrod-guided rocker arm is representative of performance versions of 351C, 429 and 460 engines. Non-adjustable positive-stop stud at top is used in the late 429CJ, all 429 police and pre-'75 460 police engines. Drawing courtesy Ford.

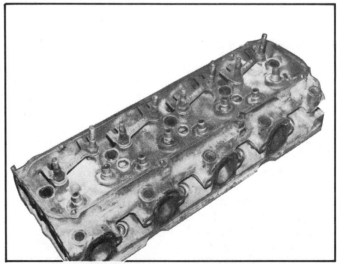

Pushrod guideplates of late 429CJ head are clamped under rocker-arm studs. Head is distinguished from 429SCJ or pre-11-1-69 CJ heads by positive-stop rocker-arm studs.

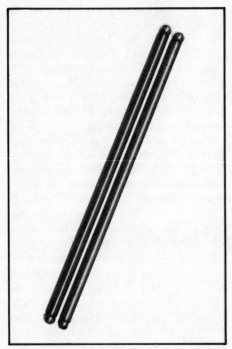

STANDARD PUSHRODS (6565)				
Engine	Year	Length (Inches)	Part Number	Note
351C	70–74	8.41	D0AZ-A	1
351C Boss	71	8.52	D0OZ-F	2
351C H.O.	72	8.41	D2ZZ-A	2
351M	75–80	9.51	D1AZ-A	1
400	71–80	9.51	D1AZ-A	1
429/460	68–69	8.70	C8VZ-A	1,3
	69–72	8.67	C9VZ-A	1,4
	73–80	8.62	D20Z-A	1
429CJ,SCJ, Police	70–72	8.55	D0OZ-B	1,2
460 Police	73–74	8.55	D0OZ-B	1,2
	75–78	8.67	C9VZ-A	1

1. 0.060 inch oversize/undersize pushrods available.
2. Hardened, must use these with guide plates.
3. Before 4-1-69, change L8.
4. From 4-1-69, change L8.

Look close to see the difference: A standard-length pushrod next to its 0.060-inch shorter cousin. If your engine has fixed rocker arms, be aware of Ford's replacement ±0.060-inch-from-standard-length pushrods.

rocker arm. Once lash is set, a *jam nut* locks the adjusting nut in position.

Pivot-guided rocker arms are not adjustable simply because the rocker-arm fulcrums are clamped to their rocker-arm stands, fixing the height of the rocker arm like the positive-stop stud, but in a different manner.

PUSHRODS

Two things should be known about 335 and 385 pushrods, one of which I've already alluded to—hardened pushrods. They must be used with pushrod-guided rocker arms, otherwise they will gall. The other item is pushrod length. Due to lack of valve adjustment with the non-adjustable rockers, something is needed to compensate for distance changes between the lifters and rocker arms after major engine work is done such as valve-seat or face grinding, cylinder-head milling or camshaft replacement. These frequently change the lifter and rocker-arm relationships more than the lifters can compensate for. If this situation occurs, the valves need "adjusting." This is accomplished with different length pushrods. Ford offers replacement pushrods which vary ±0.060 inch from their standard length. Refer to the nearby chart for pushrod lengths. Determining the correct pushrod length is covered in the assembly chapter.

CYLINDER HEADS

Cylinder-head interchangeability exists in the 335 and in the 385 engine series, but *not between* the series. In other words, 351C, 351M and 400 heads are interchangeable under certain conditions (the same applies to the 429 and 460 heads), however 351C or 351M/400 heads will not fit a 429/460 engine.

Cylinder-Head Basics—To interchange cylinder heads, you'll have to understand some cylinder-head basics. Let's start with compression ratio and how it's figured.

As a piston travels from BDC to TDC, it *sweeps* through or displaces a volume called *swept volume* (S.V.). With the piston at TDC, the volume above it is called *clearance volume* (C.V.). Clearance volume includes combustion-chamber volume in the cylinder head, volume created by the head gasket spacing the head above the block deck surface, additional volume created by piston deck clearance and the shape of the piston top. A concave or dished piston adds clearance volume. A convex or pop-up piston reduces it. Of these four factors, cylinder-head combustion-chamber volume is the biggest contributor to clearance volume.

A cylinder's compression ratio is directly proportional to its *clearance volume* and its *swept volume* or *displacement*:

$$\text{Compression ratio} = \frac{S.V.}{C.V.} + 1$$

The compression ratio formula says that *as an engine's displacement is increased, its clearance volume must increase to maintain the same compression ratio*. This is a very important consideration when interchanging cylinder heads, particularly these days when gasoline octane ratings have fallen like the value of the dollar. Excessively high compression will cause detonation problems. On the other hand, low compression reduces engine power output and mileage.

To see exactly what I mean, look at a couple of examples from the compression-ratio chart and see what happens when heads are interchanged. The '72 400, equipped with a 78.4cc (cubic centimeters) cylinder head, has an 8.4:1 compression ratio whereas the '70 351C-4V head has a 62.8cc combustion chamber

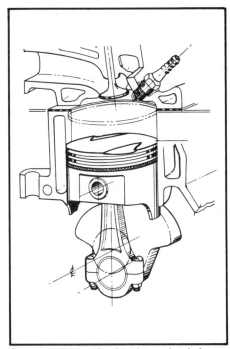

Compression ratio is determined by a piston's swept volume and clearance volume. Swept volume is displaced by piston as it travels from TDC to BDC. It is also an engine's displacement divided by its number of pistons. Volume above a piston at TDC is clearance volume.

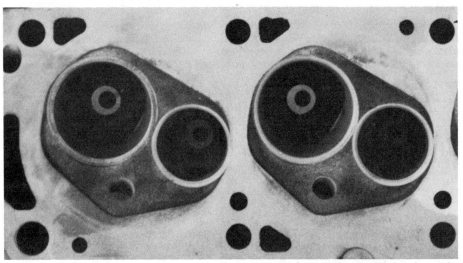

Two basic 351C combustion chambers—open and closed. Open, or large chamber, is used on the 2V, CJ and '72 and later 4V cylinder heads. Closed-chamber heads are on Boss and early 4V engines.

and a 11.0:1 compression ratio. Swapping the heads results in a 351C that has almost a 2-point lower compression ratio at 9.22:1. Its power and mileage will be lower, however the engine's octane requirements will also be lower. Put the 351C-4V head on a 400 and its compression increases to 9.6:1. This requires premium fuel, but performance increases noticeably.

Because of the swept-volume-to-clearance-volume relationship, you'll find that oversize pistons are usually dished whereas the standard piston is flat. This is because as a bore is enlarged, its swept volume increases. Consequently the clearance volume must be increased to maintain the S.V.-to-C.V. ratio to keep the original compression ratio.

Other Considerations—When interchanging heads, pushrod lengths must be matched to the engine and the rocker-arm type used. Cylinder-block deck-height is directly reflected in pushrod length, so in the case of the 351C and 400, the pushrod that goes with the engine must be used. The catch here is 351C Boss or HO heads on the 351M/400 require pushrod-guided rocker arms. Hardened pushrods are not available from Ford for the 351M/400 but can be obtained from aftermarket manufacturers.

Cylinder-Head Evolution—Now that I've covered the ins and outs of interchanging cylinder heads, let's look at the evolution of the 335 and 385 Series cylinder heads beginning with the 351C and 429, respectively.

351C-2V—This head remained basically unchanged throughout its production run. 2.04-inch and 1.66-inch intake and exhaust valves are used with 0.342-inch stems. Stamped-steel, pivot-guided rocker arms are used. Its 76.2cc combustion chamber was used with a flat-top piston to produce 9.5:1 compression ratio.

351C-4V/CJ—The '70 351C-4V cylinder head has a relatively high 11.0:1 compression ratio due to its somewhat small 62.8cc combustion chamber. Its intake and exhaust valves are larger than the 2V version at 2.19 inch and 1.71 inch. Ports are larger than the 2V head. Like the 2V, stamped-steel, pivot-guided rocker arms are used.

The '71 CJ, the only year it was produced, essentially used 4V porting and valves with the larger 2V combustion chamber. Consequently, it has less compression than the early 4V head. The CJ name was dropped after '71, but the engine was continued under the 4V name through the '74 model year, then it was discontinued.

351C Boss and HO—The '71 Boss and '72 HO heads are essentially the same with exception of combustion-chamber size. The Boss combustion chamber is smaller at 66.1cc which, when combined with its pop-up piston, produces an 11.7:1 compression ratio. The HO combustion-chamber volume is 75.4cc resulting in a 9.2:1 compression ratio with a flat-top piston. Both heads use stamped-steel, pushrod-guided, adjustable rocker arms and 4V-size ports and valves—2.19 inch and 1.71 inch.

PARTS 35

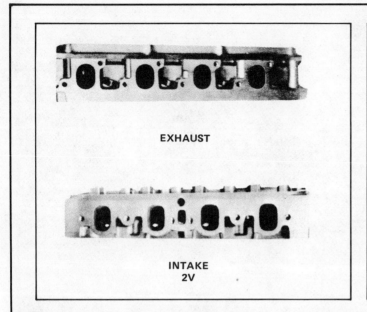
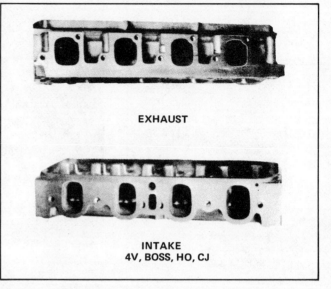

351C small- and large-port heads. Cross section of the intake and exhaust ports as viewed from intake- and exhaust-manifold faces of two heads shows the free-breathing large-port 4V, Boss, HO and CJ heads versus the smaller 2V ports.

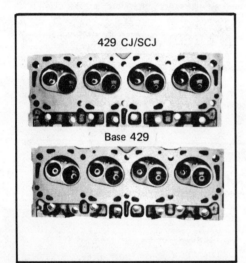

Two 429/460 cylinder heads. Top is 73.5cc small-chamber 429 CJ/SCJ with its large 2.245/1.725-inch valves. Lower photo is standard '68—'72 429/460 cylinder head with 75.8cc combustion chambers and 2.084/1.653-inch valves.

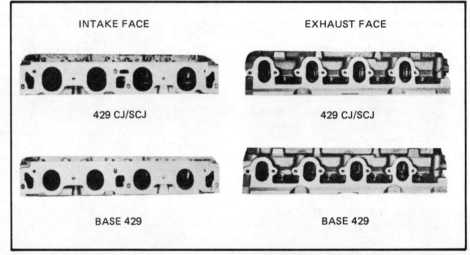

Port size difference between the base, or standard 429/460 cylinder head and the performance oriented 429 CJ/SCJ. The CJ/SCJ version has much larger intake and exhaust ports.

Two types of keepers you're likely to see, multi-groove type at left and single-groove keeper on right. Multi-groove keepers are used in all 351C and 351M/400 engines except 351C Boss and HO, and 1979 and later 351M/400 exhaust valves. Late-model 351M/400s use single-groove retainer in conjunction with a two-piece spring retainer on exhaust valves only. 429/460 uses single-groove keeper and one-piece spring retainer exclusively. Drawings courtesy Ford.

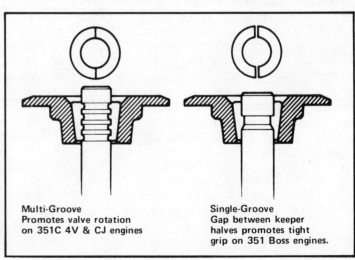

Multi-Groove
Promotes valve rotation on 351C 4V & CJ engines

Single-Groove
Gap between keeper halves promotes tight grip on 351 Boss engines.

429/460—Standard 429 and 460 cylinder heads are essentially the same on a year-to-year basis. As a result, they are 100-percent interchangable. '68–'71 heads use the cast-iron, rail-type rocker arms and positive-stop studs. Chamber volume is 75.8cc. Valve sizes remained the same at 2.084 inches and 1.653 inches throughout the 429/460's production. Pivot-guided, stamped-steel rocker arms replaced the rail rockers in 1972, and the combustion-chamber size was increased to 91.4cc, making the 429/460 a low-compression emissions engine. In 1974, the 460 combustion chamber was further increased to 96.2cc as the last significant change to this cylinder head.

429CJ and SCJ—CJ heads produced prior to Nov. 1, '69 and SCJ heads are identical. They have 73.5cc combustion chambers and 2.245-inch and 1.725-inch intake and exhaust valves. Rocker arms are stamped-steel, pushrod-guided and adjustable, the same as the 351C Boss and HO engines. CJ heads built after Nov. 1, '69 use non-adjustable pushrod-guided rocker arms. Positive-stop rocker-arm studs are used. Otherwise, they are the same as the earlier heads.

429 and 460 Police—1971 429 Police cylinder heads are essentially the same as the late 429 CJ heads with non-adjustable rocker arms. Combustion-chamber sizes were increased to 91.5cc in 1972 for lower compression, otherwise the '71 and '72 heads are the same. In 1973 when the 460 became the Police engine, combustion-chamber volume was 89cc and valve sizes were reduced to 2.197 inches and 1.653 inches. Pushrod-guided stamped-steel rocker arms and positive-stop studs are retained. The last Police head, produced from 1975 on, is identical to the standard passenger-car head. It has 96.2cc combustion chambers, 2.084-inch and 1.653-inch diameter valves and non-adjustable stamped-steel pivot-guided rocker arms.

Common single-piece spring retainer shown alongside two-piece retainer, or retainer and sleeve used with '79 and later 351M/400 exhaust valves. Retainer-and-sleeve design promotes valve rotation as does multi-groove keeper.

335 SERIES CYLINDER-HEAD SPECIFICATIONS									
Engine	351C-2V	351C-4V	351C-CJ	351C-4V		351C Boss	351C H.O.	351M	400
Year	70-74	70-71	71	72-73	74	71	72	75-80	71	72-80
Clearance Volume (cc)	76.2	62.8	75.4	75.4	78.4	66.1	75.4	78.4	78.4	78.4
Compression Ratio	9.5:1	11.0:1	9.0:1	9.0:1	8.7:1	11.7:1	9.2:1	8.0:1	9.0:1	8.4:1
Valve-Head Dia. I/E (In.)	2.041/1.654	2.190/1.710	2.190/1.710	2.190/1.710	2.041/1.654	2.190/1.710	2.190/1.710	2.041/1.654	2.041/1.654	2.041/1.654
Valve-Stem Diameter (In.)	0.342	0.342	0.342	0.342	0.342	0.342	0.342	0.342	0.342	0.342

All 335 Series engines use stamped-steel rocker arms. Boss and H.O. rocker arms are adjustable. They are stabilized by guide plates which are located under screw-in rocker-arm studs.

385 SERIES CYLINDER-HEAD SPECIFICATIONS									
Engine	429		429 Police	429CJ, SCJ	460			460 Police	
Year	68-71	72-	71-72	71-72	68-71	72-73	74-	73-74	75-78
Clearance Volume (cc)	75.8	91.4	73.5	73.5	75.8	91.4	96.2	89.0	96.2
Compression Ratio	10.5:1-2V 11.0:1-4V	8.5:1	11.3:1	11.3:1	10.5:1	8.5:1	8.0:1	8.6:1	8.0:1
Valve-Head Dia. (In.) I/E	2.084/1.653	2.084/1.653	2.245/1.725	2.245/1.725	2.084/1.653	2.084/1.653	2.084/1.653	2.197/1.653	2.084/1.653
Valve-Stem Diameter (In.)	0.342	0.342	0.342	0.342	0.342	0.342	0.342	0.342	0.342
Rocker-Arm Type	Cast-Iron, Non-Adj.	Stamped-Steel, Non-Adj.	Stamped-Steel, Non-Adj.1	Stamped-Steel, Adj.1, 2	Cast-Iron, Non-Adj.	Stamped-Steel, Non-Adj.	Stamped-Steel, Non-Adj.	Stamped-Steel, Non-Adj.1	Stamped-Steel, Non-Adj.

1 Pushrod guide plates used under screw-in rocker-arm studs.
2 CJ is non-adjustable after 11-1-69.

Charts illustrate 335 and 385 series cylinder-head evolution. Much of the complexity is related to rocker-arm design changes and emission-standard compliance.

335 Series Cylinder-Head-Interchange Compression Ratio

Comb-Ch. Volume (cc's)	351C-2V 70-74	351C-4V 70-71	351C-CJ 71	351C-4V 72-73	351C-4V 74	351C Boss 71	351C H.O. 72	351M 75-80	400 71	400 72-80
76.2	9.5	9.4	8.9	8.9	8.9	10.3	9.1	8.1	9.2	8.6
62.8	9.9	**11.0**	10.3	10.3	10.3	12.3	10.6	9.2	10.4	9.6
* 75.4	9.6	9.5	**9.0**	**9.0**	9.0	10.4	**9.2**	8.2	9.2	8.6
78.4	9.3	9.2	8.7	8.7	**8.7**	10.5	8.9	**8.0**	**9.0**	**8.4**
* 66.1	10.6	10.6	9.9	9.9	9.9	**11.7**	10.2	9.0	10.0	9.3

385 Series Cylinder-Head-Interchange Compression Ratio

Comb-Ch Volume (cc's)	429 68-71 2V	429 68-71 4V	429 72	429CJ, SCJ & Police 71-72	460 68-71	460 72-73	460 74-	460 Police 73-74	460 Police 75-78
75.8	10.5	**11.0**	9.7	11.2	10.5	9.6	9.3	9.5	9.3
91.4	9.1	9.5	**8.5**	9.7	9.2	**8.5**	8.3	8.4	8.3
* 73.5	10.7	11.3	9.8	**11.3**	10.7	9.7	9.4	9.7	9.4
96.2	8.8	9.1	8.2	9.3	8.9	8.2	**8.0**	8.2	8.0
* 89.0	9.3	9.7	8.7	9.9	9.4	8.6	8.4	**8.6**	8.4

Numbers shown in bold-face indicate standard compression ratios.
* Hardened pushrods required.

Assuming all mechanical components are compatible, never interchange cylinder heads without first checking compression ratio. You could end up with an engine that won't run on available gasoline due to excessively high compression or one that performs like a dog because of low compression. Use this chart for reference.

4 | Teardown

Tearing down your engine is the first rebuilding step. What you find through close inspection of your engine's components will "tell" you which parts have to be replaced, or to what extent they must be reconditioned. It also gives you a first-hand look at how engine servicing and operation affected internal wear. Don't look at the engine teardown phase as merely involving getting your engine apart.

REMOVE EXTERNAL HARDWARE

While your engine is still "on the hook" after being lifted out of its engine compartment, remove as much external hardware as possible before setting it down. It will be more convenient to do it now, rather than after it's on the floor or a workbench. Another tip: Use small boxes or cans for storing fasteners and small parts. Store them in groups according to their function. For example, keep the exhaust-manifold bolts in the same container rather than in different boxes containing some oil-pan, valve-cover bolts and the like.

Mark the Distributor—Put a *scribe mark* or scratch on the distributor housing and one on the engine block to match it. Or simply remember the vacuum diaphragm points approximately straight ahead in the installed position. This provides a reference for positioning the distributor when you reinstall it. Disconnect the spark-plug and coil wires, then remove the distributor cap with the wires. Remove the distributor after removing the hold-down clamp and disconnecting the vacuum advance hose/s.

Fuel Pump—Remove the fuel-pump-to-carburetor line by disconnecting the rubber hose at the carburetor fuel filter, then unthreading the line's tube nut from the pump. The pump is secured to the side of the 351C, 351M and 400 engines' integral front cover by a bolt at the top and a stud and nut at the bottom. Remove the bolt first, then the nut to remove the pump. On 429/460 engines the pump is secured to the aluminum front cover by two bolts at the pump's sides.

Carburetor—Disconnect the heat tube from the automatic choke plus any additional wiring-harness and vacuum-tube connections that may still run to the carburetor. Label everything so you can get it back correctly during reassembly. Remove the 4 carburetor mounting nuts and lift the carburetor and its spacer off the intake manifold.

Water Pump—Remove the water-pump mounting bolts. On 429/460 engines, loosen one of the thermostat-bypass hose clamps. Finish removing the bypass hose after the pump is out of the way. Pry the pump loose from the timing chain cover with a screwdriver, but watch out for the flood! Coolant will pour out of the block when the seal is broken.

Engine Mounts—Your engine mounts should still be bolted to the block. Remove them, but keep track of the bolts by threading them back into the block and tightening them.

Pressure Plate and Disc—Before removing your clutch, if your car has a standard transmission, mark the pressure-plate cover and flywheel with a center punch so they can be replaced in the same relative position. When removing the clutch, loosen the 6 mounting bolts a couple of turns at a time in rotation. This will prevent the pressure-plate cover from being twisted as a result of being unevenly loaded. When the pressure plate appears to be loose, you can remove the bolts. Be prepared to handle about 25 pounds of pressure plate and disc. One final point, if your clutch or a portion of it can be reused, *avoid getting any grease on the friction surfaces*, particularly the disc. One greasy fingerprint can make a clutch chatter and grab.

Flywheel or Flexplate—Your engine has a flywheel if it was mated to a standard transmission and a flexplate in the case of an automatic. There's no big difference about removing either except the flywheel weighs about 30 pounds more, so be ready to handle some weight. When loosening the attaching bolts, you will probably need something to keep the crankshaft from turning. A friend with a wrench on the crankshaft-damper bolt is good for holding the crankshaft.

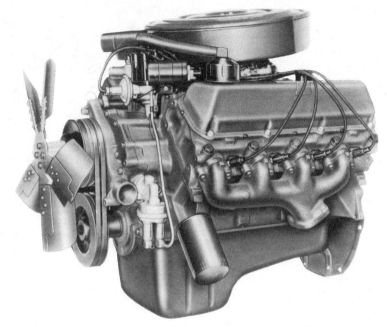

429/460 is a flexible engine as evidenced by its use in luxury and high-performance cars as well as trucks. Photo courtesy Ford.

Production history of your engine is on an aluminum tag bolted to the front of your engine, or a decal on the front of the right-hand valve cover. Information includes engine displacement, production date, and change level. Don't lose or destroy either!

Drain Oil and Coolant—If you haven't drained the crankcase oil and coolant from the block, do it now. It's the last chance you'll have to do it without creating a mess. As far as the coolant remaining in the block, the easiest way of draining it is to knock out the *freeze plugs*, or speaking more accurately, *core plugs*. Most of the coolant will run out if you knock out the lowest plug on *each side* of the engine. Use a hammer and punch to drive the plug *into* the block. Watch out for the coolant because it'll come pouring out. To remove a plug from the block once it is knocked in, pry it out with Channel-lock-type pliers.

GETTING INSIDE THE ENGINE

Now's the time to prepare to open up the "patient." Find a suitable place to work and set the engine down. A strong workbench with a work surface about 30 inches off the floor is ideal. However, if you're one of the few who own an engine stand, by all means use it. Don't feel slighted if you don't have one. Professional rebuilders usually use benches.

Exhaust Manifolds—Your engine should have a sheet-metal duct on the right exhaust manifold for heating carburetor inlet air. It'll have to come off before you attempt to remove the manifold. You may also have lifting lugs attached to both heads with the manifolds. Remove them while you're loosening the manifolds, then remove the manifolds.

INTAKE MANIFOLD

First remove the valve covers. With them out of the way you'll have better access to the intake manifold. Covers stuck to the heads can be popped loose by prying under their edges with a screwdriver after removing the attaching bolts. Work around the cover to minimize bending the flange. With the covers off, you have your first look at the inside of your engine.

After removing the intake-manifold bolts, you'll quickly discover the valve covers were an easy touch compared with the manifold. It will be stuck to both cylinder heads and the top of the cylinder block. Wedge a screwdriver between one of the corners of the manifold and a cylinder head. Once the manifold breaks loose, lift it off—it's heavy! The valve lifters and pushrods will now be exposed.

REMOVING THE CYLINDER HEADS

Rocker Arms and Pushrods—Now you are about to zero in on the real serious stuff. Loosen the rocker-arm nuts or bolts so the rockers can be rotated to the side, freeing the pushrods so they can be

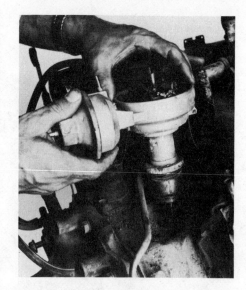
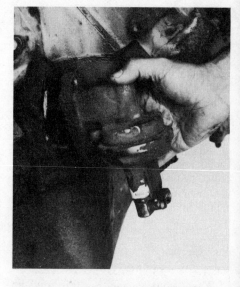

Start your engine teardown by removing external hardware. Distributor, fuel pump, carburetor and water pump are being removed here.

lifted out. As you remove the rocker arms and their fulcrums, or pivots, slip them over a wire so the mating rockers and fulcrums stay together in pairs. If they are mixed up, you can end up with scored rockers and fulcrums when the engine is run. Again, the simplest way of keeping the rockers and fulcrums together is to string them on a wire starting with cylinder 1 and proceeding to 8 in order of removal. Tie the ends of the wire together and they'll keep until you're ready for them.

Unbolting the Heads—Each head is secured with 10 bolts—count them. Five are evenly spaced along the bottom of the head and 5 more are in with the valves and springs. They are hidden from view when the valve cover is in place. I may be insulting your intelligence by suggesting you make sure *all* the head bolts are removed before you try breaking the

Breaking loose flexplate bolts with engine blocked up on the floor. Be careful when doing this so you don't roll your engine over—the bolts are tight.

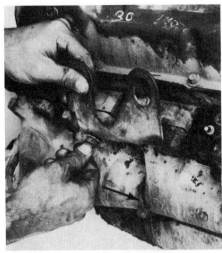
Remove engine-lifting lugs and exhaust manifolds. Notice crack (arrow). Right manifold is susceptible to cracking because it runs hot due to shrouding from carburetor heat duct. Numbers painted on valve cover are the results of a compression check. Cylinder number 3 was bad. Photo on next page reveals the reason.

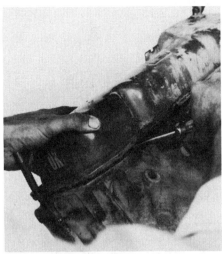
With securing bolts out carefully remove your valve covers by prying under their flanges. If you do bend one, they are easily straightened by tapping on their backsides with a hammer with valve cover on a flat surface.

This engine went too long between oil changes and was operated too cool. The engine didn't have a thermostat. Sludge forms when an engine runs too cool. Sludge is dried and flaky because of overheating. Core plugs were badly rusted and leaked. Note also incorrectly installed baffle on the rear rocker arm (arrow).

Result of the incorrectly installed rocker-arm baffle. Rocker-arm fulcrum at left shows normal wear and can be reused whereas one on right is badly galled and worn. Oil from pushrod wasn't deflected by baffle into rocker's pivot area where it was needed, but squirted over rocker arm.

Remove rocker-arm nuts or bolts, then you can remove rockers, fulcrums and pushrods. As you remove each rocker arm and fulcrum, string them *together* on a wire like I'm doing with this 429's rocker arms and fulcrums.

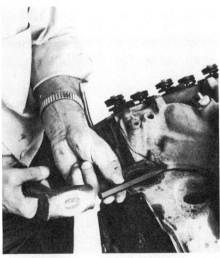
With the bolts out, break your intake manifold loose from block and heads by wedging a screwdriver or chisel between block and manifold. Just make sure *all* manifold bolts are removed, particularly if you have a 351C or 351M/400. Outboard choke-heater bolts thread into head and not manifold. Here are two bolts that weren't removed and fell victim to a long pry-bar.

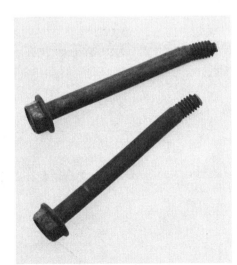
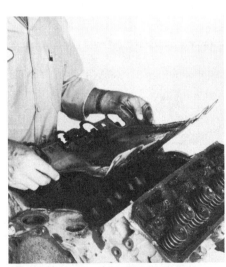
Except for early 429/460s which have a separate baffle, remove the intake-manifold baffle/gasket from top of your engine. It's under a dowel pin at front of left head and rear of right one.

TEARDOWN 41

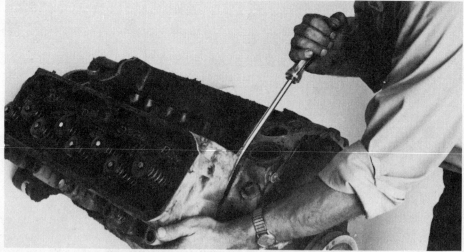

After removing all ten bolts from each cylinder head, break them loose from the block. Do this with a long pry bar inserted in an intake port as is being done with this 429, or wedge and pry between the block and head with a screwdriver as shown with this 351C. Loosely thread a couple of head bolts back in first to prevent a head from falling off when it breaks loose.

It's apparent why number-3 cylinder couldn't build compression—a burned exhaust valve (arrow).

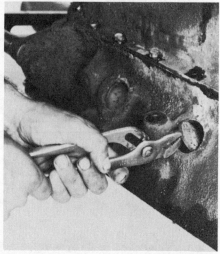

How to remove a core plug—knock it in, then pry it out. These core plugs were badly rusted and leaked as shown by rust streaks on block. Evidence of heavy coolant loss and dried sludge supports the suspicion that this engine was overheated several times. Consequently, warped heads and scuffed pistons are a couple things to check for.

heads loose from the block, but I've seen more than one enthusiastic individual attempting to pry a head loose from a cylinder block with one or more head bolts in place. Before removing the right head from a 351C, 351M or a 400 engine, make sure the oil dip-stick tube is unbolted from it.

After my cylinder-head sermon, I'll now suggest you run a couple of bolts back in each head so about two threads are engaged. This will prevent a head from falling off the block and onto the floor. This could damage a head as well as a foot. After the heads are loose, remove the bolts and lift them off.

Cylinder heads can be stubborn to get off. After the bond between the head and the block is broken, the job is relatively easy. To break the bond, you can use one of two methods. First, insert a bar, a breaker is perfect, into one of the intake ports and lever up on the head. If that doesn't break it loose, wedge between the head and block with a sharp chisel or screwdriver. Once the heads are broken loose, you can remove the "safety bolts" and lift the heads off. Your engine will now be more than 100 lbs. lighter. Mark the cylinder heads and head gaskets RIGHT and LEFT to assist in identifying problems.

NOW FOR THE BOTTOM END

To remove the crankshaft harmonic balancer, or damper, you'll need a puller. If you haven't already done so, remove the crankshaft drive pulley, then remove the center attaching bolt and washer. Now, just because the bolt is out, it doesn't mean the damper is going to slide right off the crankshaft. Quite the contrary, this is where you'll need the puller. *Don't try prying the damper off.* The

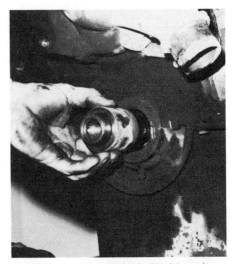

You'll need a puller to remove the crankshaft damper. Remove the damper attaching bolt and washer, then reinstall the bolt without its washer so the puller will have something to push against. Damper will clear bolt head as it's pulled off. 429/460 is equipped with a damper spacer. Slide it off end of crank after damper is out of the way.

only thing you'll accomplish is ruining the damper and damaging the front cover or oil pan.

You'll need a puller to get your damper off. Thread the damper mounting bolt without its washer into the end of the crank. This gives the puller something to push against, and the ID (inside diameter) of the damper will clear the bolt head as it is being pulled off. Use a couple of crankshaft-pulley bolts to secure the puller to the damper, then pull the damper off. If you have a 429/460 engine, there will be a spacer sleeve behind the damper. It should slide easily off the end of the crank.

Roll the Engine Over—With the damper off, remove the oil pan. I do this before the front cover because of the way they fit together. You'll need a 7/16-inch socket for all the pan bolts except those next to the front and rear crankshaft seals. These require a 1/2-inch socket. Pop the pan loose by wedging under its flange, taking care to do this in several places so the flange is not severely distorted.

Oil-Pump Assembly—Remove the oil pump. It's susceptible to damage now that it's exposed, particularly the pickup. After loosening the pump, remove it and its drive shaft which inserts into the base of the distributor. The oil-pump drive shaft is one item which must be replaced. Reused ones have been known to fail, resulting in spun bearings and other unpleasant things. Don't discard it, however. Save it to compare its length and hex size with the new one. It'll also come in handy if you intend to rebuild your distributor.

Front Engine Cover—A major difference between the 351C, 351M and 400 engines

A speed handle is handy for removing oil-pan bolts. Bolts adjacent to front and rear main-bearing caps have a 1/2-inch hex and rest are 7/16 inch. With the bolts out, remove oil pan as you did valve covers.

With oil pan out of the way, remove oil pump and shaft. If the drive sides of the shaft are worn more than just shiny, replace it. Even though you plan to trash the shaft, save it—it may come in handy later.

and the 429/460 engines is their front cover, or timing-chain cover designs. Because the 351C, 351M and 400 covers are cast integrally with the engine block, with exception of the front of the cover, you'll only have to remove the flat sheet-metal plate. 429/460 front covers are cast aluminum.

To remove either type of cover, start with the bolts. The timing pointer will come loose in the case of the sheet-metal cover, but can stay with the aluminum cover. One of the pointer's ends is attached directly to the cover. Tap the back side of the cover with a hammer to pop it loose—use the butt-end of your hammer handle for the aluminum cover. Also, watch that the sheet-metal cover doesn't hang up on its two locating dowel pins.

Achilles Heel—Like most engines using a chain-driven camshaft, the chain and drive sprockets are a weak point in the durability department. What happens is the links of the chain wear causing the chain to *elongate*—erroneously referred to as *stretching*. The chain doesn't stretch. The result is the cam *retards*, or is *behind* the crankshaft in timing, causing a loss of *low-end* power. Even worse, the chain can jump a few "cogs," preventing the engine from running.

As for the cam sprocket, the OEM-installed nylon-jacketed-aluminum-type sprockets tend to split and crack, particularly in hot climates such as the southwest U.S. Sometimes, the teeth will break off. Your engine may not be equipped with this type sprocket, it may have a replacement cast-iron sprocket simply because the original sprocket was replaced earlier in your engine's life.

Prior to removing the chain and its sprockets, check them for wear if they've had less than 50,000 miles of use. The chain should not deflect more than 1/2 inch. If the chain and cam sprocket

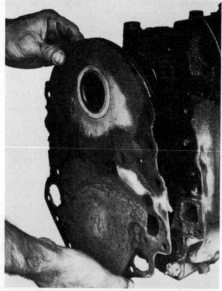
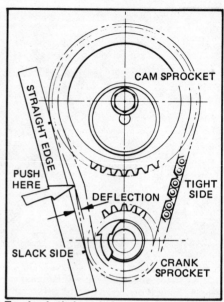

After removing bolts and timing pointer, pop timing-chain cover loose from block. 351C or 351M/400 sheet-metal front cover is shown at right and a cast aluminum 429/460 cover is shown at left. Two dowel pins at front of block locate the sheet-metal cover and can cause it to hang up and bend if you're not careful.

To check timing chain, turn crankshaft to tighten one side of chain and measure slack side. If it deflects more than 1/2 inch, replace it.

After removing fuel-pump cam, its bolt and crankshaft oil slinger, pry camshaft sprocket off. Work sprocket loose with a couple of screwdrivers like this, then slide the whole assembly off crankshaft.

have been in the engine for more than 50,000 miles, don't bother checking—*replace them*.

To check the chain, turn your crankshaft so one side of the chain is fully tight and the other side is slack. Lay a 12-inch rule or any straight edge against the chain as shown in the sketch, then push in the middle of the unsupported length of the chain. With the straight edge for reference, your chain should deflect no more than 1/2 inch. If it deflects more, replace chain *and* cam sprocket. If it has less, keep one point in mind, your chain is going to start wearing the instant you have your engine back together and running, so the less it deflects now, the longer it will last.

Getting the Chain Off—Remove the bolt and washer which hold the cam-drive sprocket and the fuel-pump cam to the camshaft. If you haven't already removed it, slide the oil slinger off the end of the crankshaft.

To remove the timing chain and sprockets, you'll have to pull them part way off together. Start the cam sprocket moving by first prying against the backside of the sprocket on opposite sides with two screwdrivers. This is a bit difficult with the 351C, 351M and 400 engines because of the way the integral front cover shrouds the cam drive. You'll have to get behind the cam sprocket from the bottom of the engine. Work the sprocket loose. Don't use too much force, particularly with the aluminum/nylon-type sprocket. Don't be surprised if the sprocket is a little stubborn. They usually are. This brings up a potential problem because the nylon teeth are easily damaged. So be careful. If the nylon has become brittle from the heat, there's a good chance it will crack regardless of how careful you are. But in this case, it should be replaced anyway.

When you get the sprocket loose, move it forward off the end of the camshaft with the crank sprocket and chain. If you don't move them together, you won't be able to get the cam sprocket all the way off without binding unless your chain is so far gone that you can slip it over one of the sprockets. Remove the cam sprocket and chain, and then the crankshaft sprocket.

After you have the chain and sprockets off, check the nylon teeth for cracks or breakage. If there are any, replace that sprocket. Also, treat it like a chain. Replace the cam sprocket if it has more than 50,000 miles on it *regardless* of its visual condition. This is particularly true if you live in a hot climate and your engine is a '73 or later emission-controlled model which operates at higher temperatures. In this case, I suggest replacing your aluminum/nylon cam sprocket with the cast-iron type.

As for the cast-iron type sprocket, they won't be cracked, but the plates which make up a *silent chain* may have worn ridges in the sprocket teeth so one or both of the sprockets should be replaced. To judge tooth wear is difficult. Drag your fingernail across the face of a tooth. It if is rough to the point of making your fingernail hang up or jump from ridge to ridge, replace the sprocket/s.

REMOVE CRANK, RODS & PISTONS BEFORE THE CAM

A camshaft can be removed before the crankshaft and rod-and-piston assemblies are out, but it's more convenient to reverse the procedure. With the crank and rods out of the way, you'll have access to the full length of the cam from inside the block rather than just the end of it from the front of the block.

Remove the Ridge—Because the top piston ring doesn't travel all the way to the top of its bore when its piston is at TDC, there is approximately 1/4 inch of unworn bore and carbon buildup at the top of the cylinder called a *ridge*. It should be removed before you attempt to remove the rod and piston assemblies, particularly if your pistons can be saved. They can be damaged easily if forced out over the ridge. If you *know* you aren't going to be using your old pistons, they can be driven out the top of their bores using a long punch or bar which reaches up under the dome of each piston. *Don't hammer on the rods*.

To remove the ridges, you'll have to use a tool appropriately called a *ridge reamer*. This tool has a cutter mounted in a fixture which can be adjusted to fit the bore. A socket and handle or wrench from your tool box is all you need to rotate the tool to do the cutting, or reaming.

As you go from bore to bore, you'll need to rotate the crankshaft periodically to move the pistons down their bores so you'll have clearance for the ridge reamer.

Fate of most high-mileage aluminum/nylon cam sprockets: cracked and broken teeth. Hot climates and emission controls cause an engine to operate at higher temperatures and aggravate the problem.

When cutting a ridge, cut the ridge *only to match the worn bore*. This is all that's required to remove a piston. Remove any more material and you may have to rebore your engine—whether it needed it or not.

Now you can get on with the rod-and-piston removal process. Roll the engine over on its back to expose the bottom end. *Before removing any connecting-rod caps,* make certain each rod and cap is numbered to match its respective cylinder. This for locating the rod-and-piston assembly in the correct bore and to assemble the rod correctly to its piston. Numbers should be on the small machined flat next to the junction of each rod and cap. Also, the numbers should be on the right side of the rods and caps that are for the right cylinder bank, and on the left side for those in the left bank. This check is a precautionary step because Ford stamps the numbers on during manufacturing. However, if they aren't numbered, you must do it. Rods and caps are machined as an assembly. Consequently, any mixup here will spell trouble.

For marking rods and caps, use a small set of numbered dies. Do the marking on the rod and cap flats adjacent to their parting line—and on the same side as their cylinder bank. If you can't obtain number dies, *prick punch* marks corresponding to the cylinder number. Or, use an electric engraver. It has the advantage that you don't have to hit the rods or caps to mark them. Mark the rods and caps before removing them from the engine.

Exaggerated section of a worn cylinder bore. Bores wear in a taper: more at top than at bottom. Short unworn section at top of a bore is the ridge, directly above top-compression-ring travel limit.

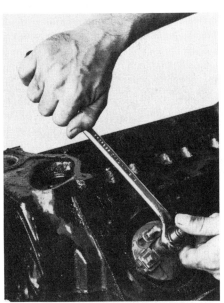

Using a ridge reamer to remove ridge at top of cylinder bore lets each piston and rod assembly slide out top of bore without the rings hanging up and causing possible ring-land damage. If you intend to reuse your pistons, doing this is essential.

Protect the Bearing Journals—After removing each rod cap, but before removing the rod-and-piston assembly, place something over the rod bolts to protect the connecting-rod-bearing journal. A bare rod bolt falling or dragging against the bearing surface during piston and rod removal can cause journal damage. Therefore, place a 2-inch piece of hose or tube over each rod bolt. This protects the

TEARDOWN 45

journals. Some parts manufacturers, suppliers and rebuilders offer flat plastic sleeves particularly for this purpose. Usually they are free for the asking—all you have to do is find the right guy to talk to. Your automotive parts store is a good place to begin.

When removing a piston, rotate the crankshaft so the piston is at BDC (bottom dead center). You'll have better access than if it were at its TDC. Anywhere in between may cause the rod to hang up between the crank journal and the bottom of its cylinder bore.

Even though the nuts are off the rod bolts, you'll find that the caps are still tight to the rod. To break a cap loose from its connecting rod, tap on the side of the rod and the cap at their parting line with a soft hammer or the end of the hammer handle. This'll do the trick. Slide the cap off the rod bolts, but have it and the nuts ready to go back on the rod.

With the rod protectors on the rod bolts, push the piston and rod assembly out the top of the cylinder. Pushing on the piston or rod with the butt end of your hammer handle and using the hammer head as a handle makes this job easy. Just make sure you are ready to catch the piston and rod as it comes out the top of its bore. Once it starts moving, it'll come out fast. Also, as you remove each rod and piston, replace the cap on the rod *with the bearings* and nuts, then set aside.

Bearings Tell a Story—Don't throw your old bearings away. They will save you the expense of farming out much of your inspection work because how a bearing wears is an accurate measure of the condition of the connecting rod. This also applies to the crankshaft. To keep the crankshaft bearings in order, tape each half together with some masking tape and record on the tape which journal the pair went with. It'll be valuable later on in the inspection and reconditioning stages.

Remove the Crankshaft—After removing the bearing-cap bolts, you'll find the main caps aren't as easy to remove as the rod caps. They fit tightly in registers machined in the block. To remove the caps, tap the end of each cap while lifting up on it and it'll pop right off.

Suppose you can remove a main cap without doing this—it's loose in its register. This indicates the cap or caps have been sprung or bent from being overloaded, usually from the engine *detonating*. Detonation is a condition where the fuel charge explodes rather than burns evenly. Sometimes detonation is referred

Before removing any caps from their connecting rods, make sure they are numbered like this. Numbers should correspond to bore. If for some reason a rod *and* cap is not numbered, do it. I'm stamping a rod and cap here with a number die *before* loosening the nuts.

Installing bearing-journal protectors on rod bolts before removing piston and connecting-rod assembly. These little plastic boots are made just for this purpose. A couple of short sections of plastic or rubber hose work just as well. Notice crank is turned so rod journal is at BDC.

I'm using a bar against piston underside to drive it over the ridge and out the top of its bore—OK unless you intend to reuse your pistons. Then you should remove the ridges so they'll slip out of their bores easily—and without damaging the pistons.

Front main-bearing inserts. Wear on grooved top bearing-insert half is due to the upward load imposed by the accessory-drive belts.

Even though bolts are out, the main-bearing caps should still be tight in their registers. Free them by tapping on their sides like this while lifting up on the cap.

Inspect the Bearings—Before tossing out the main bearings take a look at them; they'll tell you a story. Bearing inserts are made from plated copper-lead alloy or lead-based babbit, both on a steel backing or shell. If your engine has a lot of accessories such as power steering and air conditioning, the front top bearing should be worn more than the other bearings due to the higher vertical load imposed on the crankshaft and bearing by the drive belts. This wear is normal because of the way the crankshaft is loaded and should not be a cause for any concern. For the same reason, wear may also show up on the bottom of the center bearings, particularly at the second and third journals. If the bearings are copper-lead type, a copper color will show evenly through the tin plating. This makes it easy to distinguish wear because of the contrasting colors of the tin and copper. As for the lead-base bearings, it's more difficult to distinguish wear because of the similar colors of the bearing and backing.

You should be concerned about uneven wear from front-to-back on the total circumference of the bearing (top and bottom), scratches in the bearing surface and a wiped bearing surface. The first condition indicates the bearing journal is tapered, its diameter is not constant from one end to the other, causing uneven bearing and journal loading and uneven wear. Scratches in the bearing surface mean foreign material in the oil passed between the bearing and crankshaft journal. The usual cause of this is dirty oil and an oil filter which clogged, resulting in the filter being bypassed and the oil going through the engine unfiltered. A wiped bearing surface is usually caused by the journal not receiving adequate lubrication. This can be caused by periodic loss of oil pressure from a low oil level in the crankcase, a clogged oil passage or a malfunctioning oil pump. All these things have to be checked and remedied when the problem is found. Also, any problem you may find with the bearings means there may be damage to the crankshaft. Consequently you should pay particular attention to the crank journals which had bearing damage to see if there is corresponding damage to the crank.

Flex your muscles and lift the crankshaft straight up out of block. Front end is easy to lift, but you'll have to stick a couple of fingers in the rear of the crank to lift that end. Like most, this crankshaft is in good shape even though the engine has had a rough life.

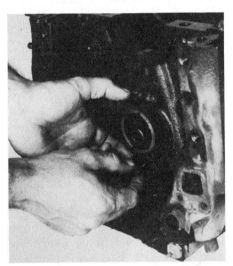 
Before removing the cam, its thrust plate will have to come off. Using cam drive sprocket loosely installed, raise lifters in their bores by rotating cam through at least one revolution. If any lobes are excessively worn, or you have a 429 or 460, you'll have to force them even higher to keep the lifters from interfering with bearing journals as you slide cam out.

If you plan on saving the cam bearings, be careful not to bump them with the camshaft while you're removing it. Use sprocket as a handle to help control camshaft movement.

If you plan on reusing your old cam and lifters, the lifters must be kept in order. As each lifter is removed from the bottom of its bore, I'm putting them in this special holder to correspond to their original position in the engine. Assuming you don't have one of these holders, a couple of egg cartons or a sheet of cardboard with 16 holes punched in it will work fine. Be sure to mark which end corresponds to the front of the engine.

Worn and new lifters. Worn lifter is useless as it is worn from its new convex spherical shape to a concave shape. Because the lifter-foot wear pattern matches cam-lobe wear, a camshaft and its lifters would quickly be destroyed if lifters are not reinstalled in their same bores.

Badly pitted and worn lobes spells the end of this camshaft. If reinstalled, it would quickly lose its lobes and spread metal particles throughout the newly rebuilt engine's oiling system.

When trying to remove oil-gallery plugs, you'll think they were put in to stay. A box-end wrench slipped over an Allen wrench should provide the additional torque needed to loosen the plugs.

to as *knocking* or *pinging.* The cause can be due to the use of lower octane-rated fuel than the engine requires or excessively high temperatures in the combustion chamber for one reason or another. If any of your bearing caps appear to be sprung, they can be reused if your engine is intended for normal street and highway use. If you're going to use it for racing, the loose caps should be replaced. The block will then require *line boring* or *honing* to true up the main-cap bearing bores with those in the block. Special bearings have to be used when this is done.

As you're removing the main caps, don't remove the bearings just yet. After you have all the caps off, lift the crankshaft straight out of the block and set it beside the block. Place the bearings and their caps alongside their corresponding journals so you can see the inserts. By doing this you'll be able to relate any bearing problems to the crankshaft journal each was fitted to. Record any potential problems you may discover and the corresponding journal so you will have this information during your crankshaft inspection and cleanup, and possible reconditioning.

Removing the Cam—Two things have to be done before you can remove the cam. First, and most obvious, is to remove the *thrust plate.* Remove the 2 thrust-plate retaining bolts, then the plate. Don't attempt to remove the cam just yet. You'll have to remove or raise the lifters in their bores so the cam lobes and bearing journals will clear them as it is moved forward.

Varnish builds up on the lifters just below the lifter bore. Mike the diameter of a lifter at this point and you'll find that its outside diameter is larger than the inside diameter of its lifter bore, making it difficult to impossible to remove up through the bore.

The best way I've found to remove a cam and its lifters is to remove the cam, then to push the lifters out the bottom of their bores. To do this, you'll have to raise the lifters in their bores and keep them there so you can remove the camshaft. *Keep* is the key word. Put the engine on its back or stand it up on its rear face. Raise the lifters by turning the camshaft a *full* revolution. Slip the cam sprocket back on to make it easier to turn the cam. If you're working with a 429/460, the camshaft lobe-lift isn't enough to raise the lifters so the cam bearing journals will clear the lifters. You'll have to force the lifters up a little higher in their bores. This would also be the case in the event your 351C or 351M/400 cam has worn lobes. Be careful not to damage the lifters if you can reuse them.

To start the cam moving forward, pry it with a screwdriver, bearing against the backside of a cam lobe and one of the bearing webs. Get ready to support the camshaft as it clears the bearings, particularly if you are doing this with the block on its back. If you are intending to save the cam bearings, be careful when doing this job. A hard cam lobe can easily damage a soft bearing if the cam is dropped.

Keep the Lifters in Order—With the cam clear of the block, you can now easily remove the lifters through the bottom of their bores. However, before you start removing them, make some arrangement to

If you are removing your cam bearings, here is one method of doing it—a mandrel and drive bar. Mandrel must fit bearing ID, shoulder against bearing shell and clear bearing bore in the block as bearing is being removed. Because all 429/460 cam bearings are same diameter, only one mandrel is required for removal or installation. You'll need at least two mandrels for the 351C or 351M/400.

keep them in order. Remember, used lifters must be installed on their original cam lobes or the lifters and the camshaft will literally be wiped out. Metal particles from the lobes and lifters will be circulated through the oiling system.

To keep your lifters in order, use anything from a couple of egg cartons to a board with 16 one-inch holes in two rows of eight. Just place the lifters in the same order as they would be in when installed in the engine. Make sure you indicate which is the front of the engine on the device you are using to keep them in order.

Now, all you have to do is push the lifters down through their bores. They'll fall right out because the lifters travel down into their bores, thus keeping the upper part of their bodies free from varnish. When pushing out a lifter, make sure you have one hand underneath ready to catch it as it falls out to prevent damage. On the other hand, if you don't plan to reuse them, let them fall where they may.

Plugs, Cam Bearings and Things—Your engine should be free of its moving parts with the removal of the cam and lifters. What's left are the oil-gallery plugs, camshaft bearings and rear plug, the oil-filter adapter and the core plugs—if you didn't remove them as suggested earlier in the chapter.

Oil-Gallery Plugs—There are 4 oil-gallery plugs, 2 at the front of the block and 2 at the rear. You'll need a 5/16-inch hex

As with all steel coolant-related components on the 351C I'm rebuilding for this book, block-to-heater-hose tube is rusted so bad it should be replaced. If your 351C or 351M/400 has suffered from the same fate, remove it with Vise-Grip pliers. Remove water-temperature sender so it won't get damaged. 429/460 sender is located in the intake manifold.

TEARDOWN 49

wrench to remove them. It's important that your oiling system be opened up so you can get it as clean as possible, particularly if your engine shows any signs of sludge buildup.

If you have hex wrenches for your socket set, this job will be much easier. All you'll need is a breaker bar in addition to a short extension for the plug behind where the distributor shaft normally is. However, if you only have the normal set of Allen wrenches, you'll need something to apply torque to the wrench and plug. A box-end wrench slipped over the end of the hex wrench works well for this.

Camshaft Bearings—Replacement cam bearings are precision inserts which assume the correct diameter after they are driven into the block. Because the bearing bores in the block are supposed to be bored on the same center line, replacement bearings installed using conventional methods *should* yield the same alignment results as the original factory method. However, it's possible they won't, meaning that the bearings must be trued, or clearanced by align-boring or the old bluing-and-scraping technique. This requires an experienced motor machinist. So consider this while you're deciding whether or not to remove your cam bearings. I found one such out-of-line bearing in building the engines for this book.

I am explaining the cam-bearing situation now so you'll understand the importance of avoiding damage during teardown and cleaning if you decide to reuse them. Be aware that a caustic hot-tank solution, so ideal for cleaning cast iron and steel, dissolves bearing material. However, the bearings can be saved *and* the block cleaned if a spray-jet tank is used, *if the cleaning solution is non-caustic*.

Remember, camshaft bearings wear so little that a normal set could outlast two or three engines. So, with these points in mind, you'll have to decide now whether you're going to attempt to save the bearings or not.

Removing Cam Bearings—If you are entertaining thoughts of removing the cam bearings yourself, don't! Unless you have an engine machine shop or access to camshaft-bearing installation-and-removal equipment and know-how to use it, farm this one out.

Cam-bearing tools vary in sophistication from a solid *mandrel* which fits inside the bearing—you'll need at least two mandrels—to a more sophisticated

If you intend to strip your block all the way, remove the oil-filter adapter. You'll need a big socket—1-1/4 inch to be exact. Shove socket against block to prevent its slippng off and rounding hex corners while applying torque to adapter.

collet mandrel which expands to fit the bearing ID. Both types have a shoulder which bears evenly against the edge of the bearing shell. The bearings are removed by a drive bar which centers in the mandrel, or by a long threaded rod which pulls the mandrel.

Either type of cam-bearing installation-and-removal tool works fine. The important thing is to make sure the bearings are installed square in their bores and without damage. I'll take up bearing installation when it's time to do so. So, if you elect to remove your cam bearings, do it now. Remove the rear cam plug, then the bearings. Drive the plug out using a long bar from the front of the block.

Oil-Filter Adapter—With the removal of the oil-filter adapter, your engine will be as bare as it's going to be with exception of the grease, dirt and old gaskets. To remove the adapter, you'll need a *big* socket—1-1/4-inch to be exact. Don't worry about it if you don't have this socket. I remove the adapter, not because it's necessary, but to remove any risk of damaging its threads.

If you do have the socket, be careful when attempting to break the adapter loose. The chamfer inside the front of the socket won't allow full engagement with the adapter's hex. Consequently, if your socket slips once, the corners will be rounded off and you can just forget about removing the adapter. To break the adapter loose, push on the socket to prevent it from slipping off the hex. If you are going to do a lot of work on these engines, grind the socket end to eliminate the chamfer, thereby making adapter removal super easy.

With the removal of the oil-filter adapter, your engine block should be completely bare and ready for inspection and reconditioning.

5 | Inspecting and Reconditioning the Shortblock

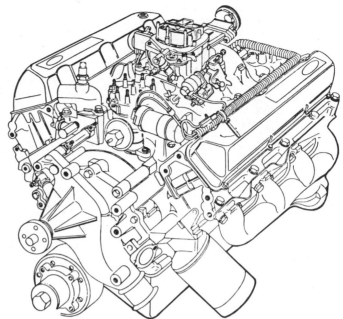

429/460 is similar in appearance to small-block Boss 302, however it is considerably larger. Major design difference between this engine series and 335 Series is separate front cover on 429/460. Drawing courtesy Ford.

Most people have an inherent knack for taking things apart, so I'll assume you've had no real problems to this point. Problems usually occur during the inspection, reconditioning and assembly processes, but don't show up until the engine is back in and running—or not running. More often than not, goofs are due to insufficient information which translates into a lack of knowledge. I've tried to include all the information you will need. Special attention has been paid to include details that are hard to get anywhere else. Therefore, if you apply this knowledge with an abundance of common sense and a reasonable amount of care, your rebuilt engine should perform better than when it was new.

CYLINDER-BLOCK CLEANING & INSPECTION

One of the most important jobs is to clean each component that will go back into your engine. After you've cleaned everything so thoroughly you are sick of it, you'll have to *keep it clean*.

Getting It Clean—If for no other reason, the size and complexity of an engine block makes it the most difficult component to clean. The job must be done right because the block is the basic foundation for supporting, cooling and lubricating all the other components. Consequently, if the cleaning job is not complete one or all of these functions will be compromised. *Get it clean!*

Before destroying the evidence, inspect the head gaskets for leaks—both compression/combustion and coolant. They will show up as rust streaks on the cylinder-head or cylinder-block mating surfaces if it's a coolant leak, or black or gray streaks radiating from a combustion chamber or cylinder if it is a combustion leak. If you find a leak connecting a cylinder with a water passage, your engine was probably losing coolant because the cooling system was over-pressurized. On the other hand, a compression leak vented to the atmosphere or another cylinder will have shown up only if you performed a compression check *prior* to tearing it down. If you discover what looks like a leak, check the block and head surfaces for warpage or other imperfections at the corresponding location after you have them well cleaned.

One type of leak to be particularly watchful for is one connecting two cylinders. The block or head may be *notched*. Notching occurs when hot exhaust gases remove metal much in the same manner as an acetylene torch. Although it is more prevalent with racing engines, it can also happen when any engine is operated for a long time before the problem is corrected. Check the severity of a notch if one exists after cleaning the cylinder head or block gasket surface.

Boil It, Spray It or Hose It Off—The cleaning process can be handled several ways. The best way is to take the block to your local engine rebuilder for *hot tanking*. This involves boiling the block for several hours in a solution of caustic soda—the longer it's in, the cleaner it gets. Get the crankshaft, heads and all the other *cast-iron and steel parts* done too. Throw in the nuts, bolts and washers, but *don't include* any aluminum, pot-metal, plastic or rubber materials you want to see again because, like cam bearings, the hot tank will dissolve or ruin them!

Now, if you would like to reuse the cam bearings, find a rebuilder who uses a *non-caustic* spray degreaser. Foreign car engine rebuilders often use this type of cleaner because these engines use a lot of aluminum. Other techniques you can employ are steam cleaning, engine cleaner and the local car wash, or a garden hose with cleaner, detergent and a scrub brush.

Regardless of which method you use, concentrate on the block interior, particularly the oil galleries and bolt holes. Use a rag with a wire to drag it through the galleries. Resort to round brushes, pipe cleaners or whatever, but pay particular attention to the oiling system. Many supermarkets carry nylon coffee-pot brushes

Lowering block into hot tank for a thorough cleaning. Caustic hot-tank solution dissolves cam bearings, and bearing material ruins hot-tank solutions. So remove your cam bearings before the cleaning-up process if you intend to replace your cam bearings.

BLOCK RECONDITIONING 51

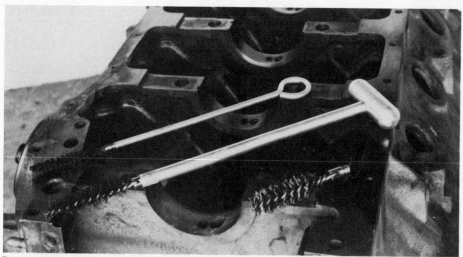

Pay particular attention to the oil galleries. Copper-bristle gun-bore brushes and high-pressure steam do the best job.

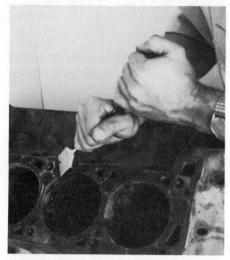

Someone has to do it. A gasket scraper makes this job much easier. Regardless of how you do it, make sure all gasket surfaces are free of old gasket material and block-to-head water passages are opened up.

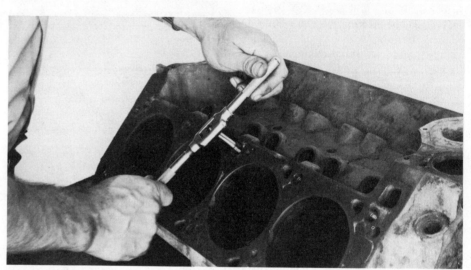

Pay particular attention to details when cleaning your engine components. Because of the critical nature of main-bearing and cylinder-head bolt torque, chase all threaded holes in the block with a tap.

which are well suited to this job. Gun brushes work well too. Team these with Ford's Carburetor and Combustion Chamber Cleaner and you'll be able to do an excellent cleaning job.

Scrape the Gaskets—Get the worst job out of the way by scraping all the gasket surfaces. A gasket scraper is a tool specifically designed for this. Now's the time to invest in one. If you try the job with a putty knife or screwdriver and then switch to a gasket scraper, you'll find out what I mean. You could've saved skinned knuckles, bad temper and loads of time.

Surfaces which need scraping are the two cylinder decks—these are the toughest—the front-cover, oil-pan, intake-manifold and oil-pump gasket surfaces. Don't stop with the block. While you're at it, scrape the heads and intake manifold too.

Don't Forget the Threads—After you think you've gotten the block as clean as it's going to get, *chase* the threads. To chase a thread, you simply run a tap in the thread to clean it out as opposed to threading it. A *bottom tap* as opposed to a *taper tap* used to start a thread, is the best type to use. You'll be shocked at the amount of

crud a tap extracts from the threads, particularly after the "spic-and-span" job you gave the block. This procedure is particularly important for threaded holes used with bolts which must be torqued accurately during assembly such as the main-bearing and head bolts. For the main-bearing threads, you'll need a 1/2-13 tap (1/2-inch diameter, 13 threads per inch) for all engines, plus a 3/8-16 tap for the four-bolt-main 351CJ and Boss 351 engines and a 7/16-14 for the four-bolt 429CJs and Super CJs. Head-bolt threads are 1/2-13 for 351Cs, 351Ms and 400s and 9/16-12 for the 429/460s.

After you've cleaned out the head- and main-bearing-bolt threads, go after the water passages. Remove any loose rust, deposits and core sand. Pay particular attention to the passages connecting the cylinder heads to the block to ensure good coolant flow to the heads. A round or rat-tail file works well for this job, but be careful of the head-gasket surfaces. A gouge in the wrong place can cause a head-gasket leak. Give the same treatment to the cylinder heads.

During all this scrubbing, scraping and general clean-up, compressed air for forcing dirt out of hard-to-get-at areas and for drying the block will be a definite help. Controlling moisture becomes more of a problem as you recondition more and more of the parts. Bearing-bore surfaces, cylinder bores, valve seats and any other machined surface will rust from humidity in the air. So, prevent this by coating the machined surfaces with a water-dispersant oil after cleaning or machining. Several brands are available from your local store, such as WD-40 and CRC. They'll do the job with a lot less fuss and mess than a squirt can of motor oil. Whatever you use, don't leave any freshly machined surface unoiled or it will rust for sure.

CYLINDER-BLOCK FINAL INSPECTION & RECONDITIONING

Inspecting your engine block to determine what must be done to restore it to tip-top condition is your first reconditioning step. To perform a satisfactory inspection job, you'll need 3—4-inch outside and inside micrometers, a *very* straight edge, and feeler gages. A set of telescoping gages will eliminate the need for the inside mikes. You may not need the straight edge if the head-gasket checked out OK. If the old gaskets didn't leak, the new ones won't either, if they are installed correctly.

Checking Bore Wear—Cylinder-bore wear dictates whether your block needs boring or just honing. This, in turn, largely determines whether you have to install new pistons—no small investment.

You can check bore wear three ways. The best is with a dial-bore gage, but you may not have one, so let's look at the remaining methods. Next in order of accuracy is the inside micrometer or telescopic gage and an outside mike. The last method involves using a piston ring and feeler gages to compare end-gaps at different positions in the bore. All of these methods will tell you what each bore's *taper* is.

Bore Taper—Cylinder walls don't wear the same from top to bottom. A bore wears more at the top, with wear decreasing toward the bottom. Virtually no wear occurs in the lower portion of a bore. Load exerted by the top compression ring against its cylinder wall during the power stroke decreases rapidly as the piston travels down from the top of its stroke. This varying load is the major cause of bore taper. In addition, the bottom of a bore which has the main function of stabilizing the piston is better lubricated and receives little wear. This is shown by the shiny upper part of a bore—while the bottom retains its original cross-hatch pattern.

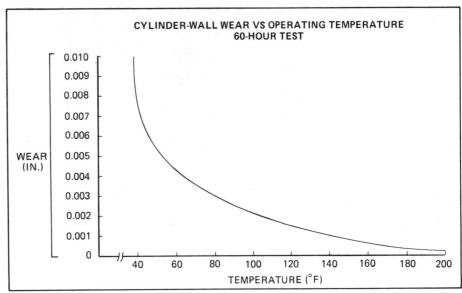

Why you should always use a thermostat. Bore wear increases dramatically as engine operating temperature goes down, particularly when it's less than 180°F (82°C). Data courtesy Continental Motors.

Measuring Taper—Because a bore wears little at its bottom, if you compare the distance across the bore at the bottom versus that at the top, just below the ridge, you can determine its taper. Also, bores don't wear evenly all the way around, nor do all cylinders wear the same. Therefore, when measuring a bore, measure it parallel to the centerline of the engine, then at 90° to the centerline. Take a couple of measurements in between and use the highest figure. It will determine what and how much must be done to that bore to restore it. Because it's not practical to treat each cylinder separately, you'll have to pinpoint the one with the worst taper and let it be the gage of what must be done to the remaining seven cylinders. One exception is when one cylinder is damaged or worn beyond the point where it can't be restored by boring, and the other cylinders are OK. In this case, it may be less expensive to have the cylinder *sleeved* rather than junking your block and buying another one.

When checking bore wear you'll notice the end cylinders have the most wear. The reason is the end cylinders operate cooler than the others, causing more wear. Most wear is concentrated on the side of the cylinder walls closest to the ends of the cylinder block in cylinders 1, 4, 5 and 8. A quick way to verify what I've said is to compare the ridges in each bore by feeling them with your finger tip, particularly the variation in thickness of the ridges in the end cylinders. Because of the wear pattern in the end cylinders, wear measured parallel to the engine centerline will exceed that measured 90° to the centerline.

Due to uneven bore wear, taper, or the difference between maximum and minimum bore wear, measuring may not provide a final figure as to how much a cylinder must be bored to clean it up—to expose new metal the full length of a bore. The reason is, uneven wear shifts a bore's centerline in the direction of the wear.

To restore a cylinder to its original centerline usually requires removing more metal than indicated by its taper. As a result, final bore-size determination is made at the time of boring. If a cylinder bore does not clean up at 0.010 inch oversize, the machinist has to bore to

BLOCK RECONDITIONING 53

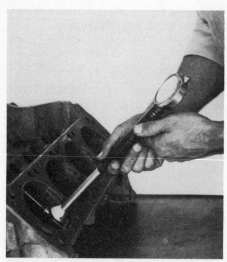  

Two direct methods of measuring bore wear, with a dial-bore gage and a snap or telescoping gage and micrometers. Worn and unworn section of each bore are measured, then compared to find bore wear, or taper.

TAPER VS RING END-GAP DIFFERENCE

$$\text{TAPER} = \frac{G_2 - G_1}{3.14}$$

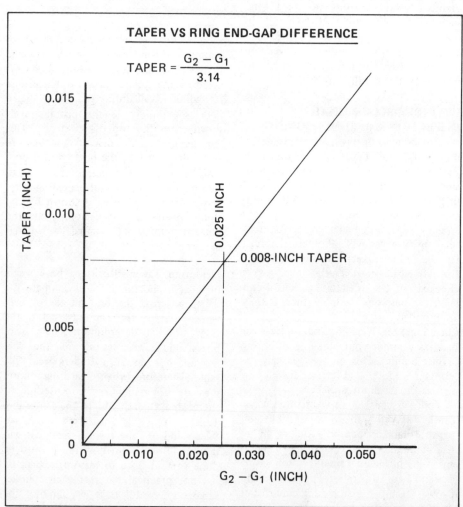

0.025 INCH — 0.008-INCH TAPER

G_2-G_1 ΔG	TAPER
0.000	0.0000
0.001	0.0003
0.005	0.0016
0.010	0.0032
0.015	0.0048
0.020	0.0064
0.025	0.0080
0.030	0.0095
0.035	0.0111
0.040	0.0127
0.045	0.0143
0.050	0.0159

Approximate Taper = $0.30 \times \Delta G$

Using ring-end-gap method of determining bore taper. Maximum end gap is found with ring placed immediately below ridge. Minimum gap is measured with ring pushed down to unworn section of the bore. Shove ring down bore with a piston to square it in bore before measuring gap. After finding the maximum difference between two end-gap readings, use this curve or chart to find bore wear. Δ is difference between G_2 and G_1.

0.020 inch, the next available oversize. The machinist first bores the cylinder with the worst wear/taper because it will establish the maximum oversize for all the cylinders.

Dial-Bore Gage or Micrometer Method—When using a bore gage or micrometer to measure taper, measure the point of maximum wear immediately below the ridge. Because wear will be irregular, take several measurements around the bore to determine maximum wear. To determine taper, subtract the measurement at the bottom of the bore from that at the top.

Ring-and-Feeler Method—Measuring bore wear with a piston ring and feeler gage is going about it indirectly. When using this method, you are actually comparing the difference between the circumferences of the worn and unworn bore. The accuracy of this method is less the more irregularity a cylinder is worn, however it is accurate enough to determine whether you'll need to bore and install oversize pistons or clean your pistons, hone the bores and install *moly rings*.

To use the ring-and-feeler method, place a ring in the cylinder and compare the difference in ring end gaps with the ring placed at the bottom of the bore and with it at the top of the bore—immediately below the ridge. Use the same ring. While it doesn't need to be a new one, it must be square in the bore to get an accurate reading. To square the ring, push it down the cylinder with a bare piston— no rings—to where you want to check ring gap. Measure the gap with feeler gages and record the results. After measuring the ring end-gap difference, use the accompanying chart or graph to determine bore taper. *Taper is approximately 0.3 times ring end-gap difference.*

How Much Taper is Permissible?—To decide how much taper your engine can live with, you'll have to ask yourself how many more *good miles* you want out of your engine. Do you want it to last 10,000, 20,000 or 100,000 miles before oil consumption takes off? If your engine has excessive taper, even new rings will quickly fatigue and quit sealing. They must expand and contract to conform to the irregular bore during every stroke. Consequently, they lose their resiliency, or springiness. Also, because ring gap must be correct *as measured at the bottom of the bore*, in a tapered bore the gap will be larger at the top where compression and combustion pressures are highest. Consequently, these pressures will be reduced slightly because of increased blowby through the wider

Here's a cylinder that's just been sleeved. All that is required to finish this sleeving job is to trim sleeve top flush with block deck surface.

Measuring piston-to-bore clearance with feeler gages. If it's 0.008 inch or more, rebore.

ring gap.

Therefore, if you just want a "Band-Aid" job on your engine so it will go another 10,000, or maybe 20,000 miles before it's right back where you started, you can *get away with* reringing a block with some cylinders having as much as 0.010-inch taper. If you are using the ring-and-feeler method, rebore if taper exceeds 0.008 inch. This is because this method masks some forms of wear. However, if your object is to have a *truly* rebuilt engine, don't merely rering if taper exceeds 0.006 measured by the dial-bore-gage or micrometer methods of checking, or 0.004 inch by the ring-and-feeler method. Remember, taper gets worse with use, never better. It's best to start a rebuilt engine's life with a straight bore and new pistons if you want maximum longevity.

SLEEVING AN ENGINE

A sleeve is basically a portable engine cylinder. It is used to replace or restore a cracked, scored or otherwise damaged bore that can't be restored by conventional boring and honing techniques. An expenditure of only $20 to $30 for sleeving a cylinder can save you the cost of a new or used engine block. A sleeve is a cast-iron cylinder which is slightly longer than the length of the cylinders in the engine it is made for. It has a smaller ID than the original bore for finishing stock and a wall thickness varying from 3/32 to 1/8 inch.

To install a sleeve, the damaged cylinder must first be bored to a size 0.001" less than the OD of the sleeve. Rather than boring all the way to the bottom of the bore, the boring machine is stopped just short to leave a *step* or *shoulder* for locating the sleeve. The step and the interference fit of the sleeve prevent it from moving around after the engine is back in operation. Some rebuilders leave more than 0.001" interference between the sleeve and the block. The problem is this overstresses the block and can distort neighboring cylinders, consequently the 0.001" figure should be adhered to.

When an engine is ready to be sleeved, it should be evenly warmed up using a torch, furnace or whatever so the block will grow. At the same time the sleeve is cooled to shrink it—a refrigerator freezer works great for this. The sleeve will *almost* drop all the way into the cylinder, however it will quickly rise to the same temperature as the block, assuming its interference fit. Then it must be pressed or driven in the rest of the way and sealed at the bottom. Some shops don't bother with heating and cooling. They drive the sleeves in all the way. The excess length of the sleeve protruding from the block is trimmed flush with the deck surface. The block is now restored and the sleeved cylinder is bored to match other cylinders.

Piston-to-Bore Clearance—All this talk about whether to bore or not to bore may turn out to be purely academic. The main reason for not wanting to bore an engine is to avoid the cost of new pistons, a legitimate reason considering they cost at least $10 to $20 each. If your old pistons are damaged or worn to the point of being unusable, you'll have to purchase new pistons whether your block needs boring or not. So, to make the final determination as to whether your engine should have a rebore, check the piston-to-bore clearances. Refer to the piston section of this chapter for other piston-related problems you'll need to identify before giving your pistons the OK.

Two methods can be used for checking piston-to-bore clearance. The first is mathematical. Measure across the cylinder bores 90° to the engine centerline and approximately 2 inches below the top of the bore. Then measure the piston *which goes with the bore*. Make it in the plane of the wrist pin and 90° to the wrist-pin axis, measuring the piston across its thrust faces. Subtract piston diameter from bore diameter to get piston-to-bore clearance. Measure with the block and pistons at the same temperature—preferably normal room temperature. If their temperatures are too far off, your measurements will not be accurate.

Direct measurement of piston-to-bore clearance is the next method. Install the piston-and-rod assembly *in its bore* without rings and in its normal position. By normal position, I mean with the notch or arrow in the piston dome pointing toward the front of the engine. Positioning and checking the piston in this manner ensures that the clearance you're checking is what the piston sees when it is correctly installed in the engine. Use your feeler gages from the top of the cylinder to check between the piston thrust face and the bore in several positions up and down the bore between the piston's TDC and BDC. The thrust side of a piston is its right skirt or the skirt on the right as the piston is viewed with its notch or arrow pointing away from you. Again, maximum clearance should not exceed 0.008 inch.

There is a method to take up the clearance between a piston and its bore, called *knurling*. Depressions are put in the pistons thrust faces creating raised projections which effectively increase diameter and reduces bore clearance. The problem is, knurling is *temporary*. The projections wear off quickly, putting the clearance right back where it was before all your work. Don't do it! Replace your pistons if their clearance exceeds 0.008 inch.

Glaze-Breaking—If the bores are within acceptable taper limits, it is good practice to *break the glaze*. Glaze-breaking doesn't remove any appreciable material from the bore, it merely restores a hone-type finish

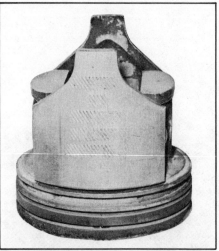

Embossed pattern rolled into this piston's skirt is a *knurl*—a temporary fix for reducing clearance between a worn bore and piston. I don't recommend it for any engine.

for positive ring break-in. If you are using plain or chrome rings, this operation is necessary. It is optional with moly rings—but desirable because of the ring-break-in aspect. Again, glaze-breaking is not done to remove material, therefore a precision-type hone should not be used because it will try to remove what bore taper is there, increasing piston-to-bore clearance. A spring-loaded or ball-type hone that *follows* the existing bore without changing its shape should be used for glaze-breaking.

DECKING YOUR BLOCK

Before sending your block off to be bored or honed, first check for any irregularities on the deck (head gasket) surfaces if either head gasket showed signs of leakage. If you have additional machining operations done on the block, you'll want to have them done *all at one time*. Another, more important reason is many engine machine shops use a boring bar which locates off the block deck surfaces. So, if the deck surface is off, the new bore will also be off relative to the crankshaft centerline.

If your block is notched between cylinders, it may have to be welded if more than the following amount has to be removed to clean the deck surfaces:
0.035 inch from 351Cs up to '71
0.0155 inch from '72 and later 351Cs
0.0155 inch from all 351Ms
0.0565 inch from 400s
0.010 inch from 429/460s up to '70-1/2
0.020 inch from '70-1/2 to '71 429/460s
0.032 inch from '72 and later 429/460s.
Removing more than the amounts indicated here will cause deck clearance problems.

Glaze-breaking, or restoring cross-hatch pattern to bores that don't require reboring is particularly important if chrome rings are to be used. This type hone follows existing bore without removing any appreciable material.

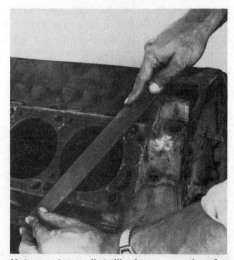

Using a large flat file in preparation for checking block deck surface for flatness. This also removes imperfections which could interfere with a boring bar which locates on the block's deck surface.

Another problem with milling a block's deck surface, particularly these days, is the engine's compression ratio will be increased, raising the engine's octane requirements. High-octane fuels are increasingly hard to find—not to mention the added cost.

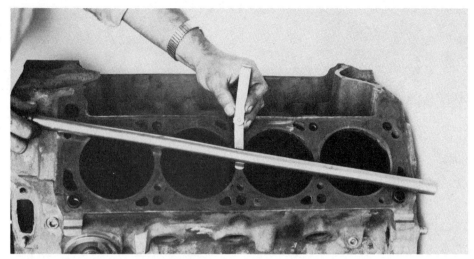

Checking deck-surface flatness with ground-steel bar and feeler gages. I prefer limiting surface irregularities to 0.004 inch, however 0.007 inch is allowed over the full length of the block.

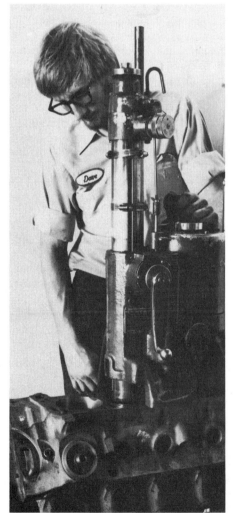

Regardless of the boring bar used, main-bearing caps should be torqued in place. Use boring bar to remove only enough material to clean up bore to accept next largest oversize piston. 0.002–0.003-inch honing stock is left.

Treat "decking" your block in the same manner as you would milling cylinder heads. If more than 0.020 inch is removed from the block, you'll also have to surface your intake manifold due to a mismatch being created between the heads and the manifold. You may also run into pushrod-length problems if your engine is equipped with non-adjustable rocker arms. Refer to pages 77-78 and 117-118 for more on this.

Because a block is cast-iron, a special welding process is required to repair it. The block must be preheated prior to welding, and a nickle-alloy rod used to do the welding. After the notch is welded, the raw weld must be machined flat, or made flush with the block's deck surface. Although it's not your problem, this is not as easy as it sounds because a nickle-alloy weld is very hard and nearly impossible to machine, consequently it must be ground. Not all automotive machine shops have block-grinding equipment. *Your* problem is the cost of having it done. Compare this cost to that of a replacement block.

For less-severe surface irregularities, check your block, particularly if there are signs of gasket leakage. You'll need an accurate straight edge at least 6-inches long and some feeler gages. Use your gages to check for any gaps between the block deck surface and the straight edge. If your straight edge extends the full length of the block, the maximum allowable gap is 0.007 inch. However, if your straight-edge is 6-inches long, the maximum is 0.003 inch. These two figures equate to basically the same amount of warpage. If you find block warpage, check back a couple paragraphs for how much can be taken off your block's deck surfaces.

A final thought about surfacing your cylinder block. If one deck surface requires surfacing, the other one must be done whether it needs it or not. This maintains equivalent deck heights from side-to-side and ensures that the intake manifold will seal.

Other Checks—Other than checking your block for cracks, bore taper and deck flatness, there's not much more to look for. In the normal course of things, nothing can go wrong with a block. For example, the valve-lifter and distributor bores are lightly loaded and so drenched in oil that they just don't wear. The only problem would occur if they were damaged by a broken connecting rod which somehow got into the cam and lifters. This would require junking the block.

To be sure the lifter bores are OK, check them with a *new* lifter. Do this without oiling the lifter or the bores. Insert the lifter in each bore and try wiggling it—if there is noticeable movement, the bore is worn and the only way of correcting it is to replace the block. Sleeving a lifter bore is possible, but difficult to have done.

If you're building a racing engine, another check you'll need to make is for crankshaft bearing-bore trueness—making sure each bore is on the same centerline. The easiest way of doing this is to spin your crank after installing it on its bearings in the block—the crank should be checked for runout first. If the crankshaft turns freely, your block is OK as is the crankshaft.

A final check for cracks and your block can be considered sound. Be particularly attentive if you have a 351M/400 cast prior to March 1, 1977 at the Michigan Casting Plant. The water jackets were prone to cracking immediately above the lifter bores. Refer back to pages 27 and 28 for a more detailed description of this problem.

CYLINDER-BORE FINISHING

Plain Cast-Iron, Chrome or Moly Rings? When you drop your block off to have it bored or honed, you should know the type of piston rings you intend to use so you can tell the machinist. Your engine's final bore finish should be different for plain, chrome or moly rings. So, what type of rings should you use?

Chrome rings are especially suited for engines that will inhale a lot of dust and dirt such as a truck used in a rock quarry.

BLOCK RECONDITIONING 57

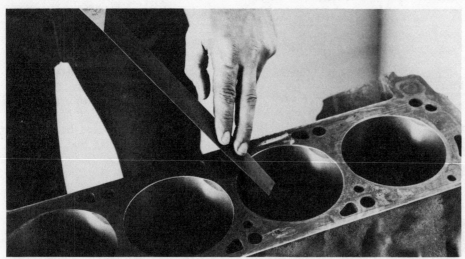

Boring leaves a sharp edge at the top of each bore. Chamfer bores with a file to provide a lead-in for installing the rings and pistons. This can be done after honing, however doing it first eliminates danger of damaging finished bore with file end.

Honing block with a hand-held hone after boring. Bore diameter is checked periodically as correct bore size is approached. It is particularly important that main-bearing caps are in place and their bolts torqued to spec to produce accurately honed bores.

Chrome is very tough and lasts longer than moly under these conditions. However, unless you have this situation, use moly rings. Their life will be much better than the chrome type, and they require virtually zero break-in mileage. As for plain cast-iron rings, they break in quicker than *chrome-faced* cast-iron rings, but they also wear out quicker. I recommend you don't consider them.

What's the Difference—The advantage of a moly ring over a plain or chrome one is the moly-type carries more of its own oil. It has surface voids—little depressions—which contain oil much in the same manner as the cross-hatching of a honed cylinder. Plain and chrome rings have virtually no voids and must depend almost totally on cylinder-wall cross-hatch to supply lubricating oil. When the piston travels down during the power stroke, the cylinder wall is exposed to the burning fuel. Consequently, the oil on the wall is partially burned away, meaning the piston rings will not receive full lubrication during the return trip up the bore on the exhaust stroke. The moly-type ring carries its own lubrication which is not directly exposed to combustion.

The reason for different bore finishes according to the type ring used should now be obvious. A chrome ring depends on a coarsely finished cylinder wall to retain lubricating oil. If plain or chrome rings are used, the bore should be finish-honed with a 280-grit stone. A 400-grit stone is used for moly rings. A 30° cross-hatch pattern is used for all ring types.

Install the Main-Bearing Caps—Prior to delivering your block for boring and/or honing, install the main-bearing caps. Torque them to specification: 95–105 ft. lbs. for all 1/2-13 main-bearing bolts. For those using four-bolt mains, the outer bolts of the 351C CJ and Boss 351 are torqued 35–40 ft. lbs. and the 429CJ and Super CJ are torqued 70–80 ft. lbs. This is necessary because the load imposed on the block by the main-bearing bolts distorts the bores slightly. Therefore, the bores are not the same shape before and after the main cap bolts are torqued. Consequently, the object of boring and honing the block with the main caps torqued to spec ensures that the bores will be as close to perfect by simulating the normal operating stresses and deflections in the block during the machining processes.

Cylinder heads are another source of bore distortion, but how can you hone the engine with the heads in place? Fear not, the dilemma has been solved. Some engine shops use a *torque plate* when boring and honing. The 2-inch-thick steel torque plate has four large-diameter holes centered on the engine bores to allow clearance for boring and honing. The plate is torqued to the block just like the heads. If the shop you choose does not use a torque plate, don't be overly concerned. This is not absolutely essential unless you are building an engine for all-out racing. Using a torque plate increases the machining costs due to the additional time required. Therefore, if you are on a tight budget, make some cost comparisons. Regardless of whether or not a torque plate is used, *do* install the main-bearing caps—that's free if you do it.

Chamfer the Bores—After your block is back from the machine shop, inspect the bore tops. The machinist should have filed or ground a small chamfer or bevel at the top of each bore after honing. This eases piston installation by providing a lead-in for the piston and rings, and it eliminates sharp edges that will get hot in the combustion chamber. A 60°, 1/16-inch-wide chamfer is sufficient. A fine-tooth *half-round* or *rat-tail* file works well for this job. Just hold the file at a 60° angle to the deck surface, or 30° to the bore as you work your way around the top of each cylinder. Don't hit the opposite side of the cylinder wall with the end of the file. You don't want to gouge your freshly honed bore.

Clean it Again—Even though your block was hot-tanked, spray-jetted or whatever method you used for cleaning it, your block must be cleaned again. If you are trying to save your cam bearings, inform your machinist, otherwise he'll automatically clean the block after he's machined it. So, if he uses a hot tank, it's good-bye bearings. This second cleaning removes machining residue, mainly dust and grit left from the honing stones. If it isn't removed before the engine is assembled, your engine will eat up a set of rings so fast it'll make your wallet ache. Grit will be circulated through your engine's oiling system and end up embedded in the crankshaft, connecting-rod and camshaft bearings, turning them into little grinding stones.

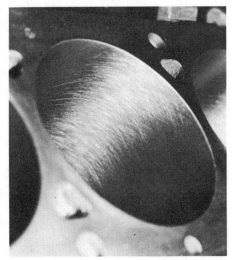
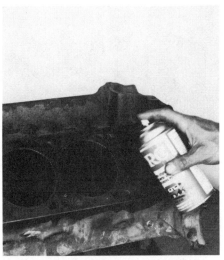

Cross-hatch pattern of a freshly honed block. The block must receive a final cleaning to remove the abrasive grit and metal cuttings. A scrub brush used vigorously with laundry detergent and hot water works well. After the bores are cleaned and rinsed, they should be dried and oiled immediately.

Consequently, even though your engine was hot-tanked by the machinist, enlist the use of a stiff-bristle brush, laundry detergent and warm water to satisfy yourself that the engine is *clean*. This, combined with a generous amount of elbow grease, will give you a super-clean block. Finish by wiping the bores with a white paper towel. It should stay clean and white if you've gotten the bores clean. If not, clean them again.

After you've finished the scrubbing and rinsing, *don't let the block air-dry*. The newly machined surfaces will rust very quickly. If you have compressed air, use it to blow-dry the block. It's best because you don't have to touch the block and the compressed air will blow water out of every little corner. The next best thing to use is paper towels—not cloth rags or towels. They spread lint that won't dissolve like that from a paper towel. *Immediately* after you've gotten the block dried off, get your spray can of oil and coat all the machined surfaces. CRC and WD-40 displace water rather than trapping it. This will protect the block against rust, but will also turn it into a big dust collector. Consequently, you'll have to keep it covered. Now that you're finished with the block until assembly time, store it by standing it on its rear face. Cover it with a plastic trash bag until you're ready for final engine assembly.

CRANKSHAFT

The crankshaft is the main moving part in any engine. Most of the other parts turn the crankshaft, support it and lubricate it. Because of its importance, a crankshaft is a very tough, high-quality component. As a result, it's rare that a crankshaft cannot be reused for a rebuild. They very rarely ever wear to the point of having to be replaced. What usually happens is the journals are damaged from a lack of lubrication, but mostly from *very dirty* oil. Usually this damage can be repaired. Even dirty oil usually doesn't change a crank's journals, but if the dirt particles get too big and numerous, it will. The one thing that can render a crankshaft useless is mechanical damage caused when another component in an engine breaks—such as a rod bolt. If an engine isn't shut off immediately or if this occurs at high RPM, the crank and many other expensive components will have to be replaced also.

Main and Connecting-Rod Journals—The most important crankshaft-inspection job is to make sure the bearing journals are round, their diameters don't vary over their length and their surfaces are free of imperfections such as cracks. A cracked crankshaft *must* be replaced. All the engines covered in this book have cast-iron cranks, however the high-performance crankshafts are stronger due to a higher *nodularity* content. So, if you've found a crack, consider yourself lucky. A crack in a cast-iron crankshaft is followed very quickly by a break. So don't even think about trying to save your crankshaft if it is cracked.

The best way of checking for cracks is by Magnafluxing, the next is by dye testing and the last is by visual inspection. Check with your local parts store or engine machine shop about having your crank crack-tested, preferably by Magnaflux. The cost is minimal. Remember—a broken crank will destroy your engine.

If a bearing journal is oval- or egg-shaped it is said to be *out-of-round*. This condition is more prevalent with rod journals than with main-bearing journals. If a bearing journal's diameter varies over its length, it is *tapered*. This is like the taper of a cylinder bore except it's an outside diameter. Finally, the journals should be free from any burrs, nicks or scoring that could damage bearings. Pay special attention to the edge of each oil hole in the rod and main journals. Round these edges with a small round file if you find any sharp ones. Sharp edges protruding above the normal bearing-journal surface can cut grooves in the bearing

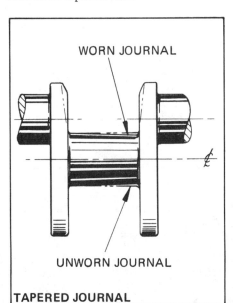
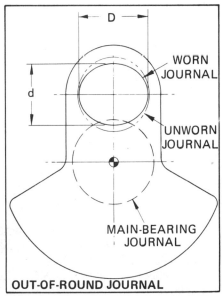

Bearing journals wear two ways, tapered or varying diameter over their length and out-of-round or oval in shape. Connecting-rod journals tend to wear out-of-round due to the way they are loaded.

BLOCK RECONDITIONING 59

Deep groove in this connecting-rod journal is from foreign material in the oil. A good polish job will make crank suitable for reuse, however more grooving would require regrinding or replacing crankshaft.

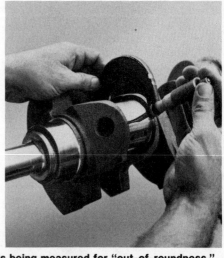

Measuring bearing journals. Front rod journal is being measured for "out-of-roundness." Record your measurements so you'll be able to compare figures.

inserts and wreck the bearing, spoiling your rebuild.

Surface Finish—Look at surface finish first. Any rough bearing journal must be refinished or reground—regardless of taper or out-of-roundness. About the only way of determining if your crankshaft journals are smooth is to run your fingernails over them. If a journal looks rough and you can feel it, new bearings would be worn quickly by this roughness. A regrinding job is then in order. A normal 0.010-inch regrind job will hopefully take care of the journal surfaces and any normal wear irregularities. I say *surfaces* because if one journal has to be ground, all rod *or* main journals should be ground the same amount. You don't want one odd-size bearing because bearings are sold in sets. The additional cost of machining all the main or rod journals is slight because the major expense is the initial setup cost. Therefore, if 0.010 inch is removed from the standard-size bearing journals, 0.010-inch *undersize* bearings will be required. Undersize refers to the crankshaft-journal diameter. The amount of undersize is the difference between the journal diameter after being reground and the nominal, or average specified standard-size journal diameter. Example: if the nominal specified diameter for main journals is 2.7488 inches and they are reground to 2.7388 inches, the crank is 0.010-inch undersize on the mains.

While you're checking bearing journals, look carefully at the thrust faces on the number-three main-bearing journal. This is particularly important if your engine is backed by a standard transmission. When a clutch is disengaged, the release bearing pushes forward on the crankshaft via the pressure plate. This load is supported by the crankshaft's rear thrust surface, so if the pressure plate had an excessively high load or your clutch was incorrectly adjusted, the thrust surface may be damaged.

An indication as to whether you'll have thrust surface trouble is the condition of your pressure-plate release levers. If they are excessively worn from the release bearing, then be particularly attentive. Measure between the thrust faces. You'll need a snap gauge and a micrometer or an old or new bearing insert, feeler gauges, and a 1—2-inch micrometer. Minimum width is 1.124 inches; maximum is 1.126 inches. If the maximum is exceeded, you'll have to replace your crankshaft or have it reconditioned. Check the thrust-face width in several places. Other dimensions and tolerances for checking your crankshaft are in the table.

Out-of-Round—To check for out-of-round or tapered journals, you need a 2—3-inch outside micrometer and an understanding of what you're looking for. If a journal is round, it will be described by a diameter you can read directly from your micrometer at any location around the journal. If it is out-of-round, you will be looking

CRANKSHAFT SPECIFICATIONS

Engine	Journal Diameter (in.)		Stroke (in.)	Runout (in.)	Taper (in./in.)	Out-of-Round (in.)	Thrust-face Width (in.)
	Main	Conn. Rod					
351C	2.7488±0.0004	2.3107±0.0004	3.50	0.004	0.0003	0.0004	1.125±0.001
351M/400	2.9998±0.0004	2.3107±0.0004	3.50/4.00	0.005	0.0006	0.0006	1.125±0.001
429/460	2.9888±0.0004	2.4996±0.0004	3.59/3.85	0.005	0.0006	0.0006	1.125±0.001
460 Police	2.9888±0.0004	2.4996±0.0004	3.85	0.005	0.0006	0.0006	1.125±0.001
429CJ	2.9888±0.0004	2.4996±0.0004	3.59	0.004	0.0003	0.0004	1.125±0.001
SCJ	2.9888±0.0004	2.4996±0.0004	3.59	0.004	0.0003	0.0004	1.125±0.001

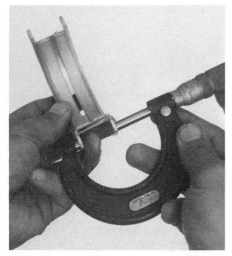
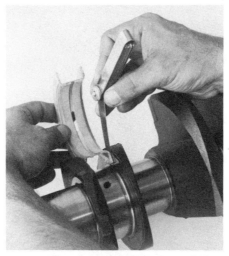

If you don't have a snap gage, here's an easy way to measure crank thrust-face width. Measure width of the bearing thrust surfaces with micrometers. Then, with the bearing fitted to the center main-bearing journal, measure clearance between bearing insert and crankshaft thrust surfaces. These two measurements add up to the crank's thrust-surface width.

Checking crankshaft runout with only the upper halves of the front and rear bearing inserts in place and oiled. Total dial-indicator reading as the crankshaft is rotated should not exceed 0.004 inch or 0.005 inch depending on the engine. Chart provides crank specifications.

for the *major and minor* dimensions of an ellipse. These can also be read directly. An out-of-round journal will be a close approximation of an ellipse. The minor dimension can be found by measuring around the journal in several locations. Major and minor dimensions will occur about 90° from each other. The difference between these two figures is the journal's out-of-round, or

Major Dimension — Minor Dimension
= Out-of-Round

Start with the connecting-rod journals when doing your checking. They are the most likely to suffer from this out-of-round condition because they are highly loaded at top and bottom dead centers. With this in mind, check each journal in at least a couple of locations fore and aft on the journal. The maximum allowable out-of-round is 0.0004 inch or 0.0006 inch, depending on your engine. Remember, if you find *one* journal out of spec, your crankshaft should be reground or replaced.

Taper—Crankshaft journal taper causes uneven bearing wear more than an out-of-round journal. This is due to uneven bearing loads over the length of the bearing. As a journal tapers, it redistributes its load from less on the worn portion or smaller diameter, to more on the less-worn part or larger diameter of the journal. This results in high bearing load and wear when a crank with excessively tapered journals is reinstalled with new bearings.

Taper is specified in so many tenths of thousandths-of-an-inch per inch. This means your micrometer readings will

Use this pattern for making a bearing-journal-radius template. A journal radius should not be larger than 0.100 inch, otherwise it will edge-ride the bearing. 1/16-inch (0.0625-inch) and 1/8-inch (0.125-inch) radii are for reference.

vary by this amount when taken one inch from each other in the same plane.

For connecting-rod journals, maximum allowable taper is 0.0003 inch/inch or 0.0006 inch/inch, as shown in the table. Remember, if one journal is not within spec, all the other journals should be reground at the same time the faulty journal is being corrected.

If your crankshaft needs regrinding, trade it in for a *crank kit*. You'll be getting someone else's crankshaft that has been reground.

A crank kit consists of a freshly reground or reconditioned crankshaft *and* new main and connecting-rod bearings. I recommend taking this approach rather than having your crank reground *and* buying new bearings. You'll end up saving about $25. Just make sure the bearings accompany the crankshaft.

After Regrinding—Check your crankshaft carefully if you've had it reground or traded yours in on a reconditioned one. Two things to look for are sharp edges around the oil holes and wrong shoulder radii at the ends of the journals. All oil holes should be chamfered, or have smooth edges to prevent bearing damage. As for the radius, there should be one all the way around each journal—where the journal meets the throw. If there's a corner rather than a radius, it will weaken the crankshaft because a corner is a weak point, or a good place for a fatigue crack to start. A radius that's too large is great for the crankshaft, but bad news for the bearings. The bearings will *edge-ridge*, or ride up the radius rather than seat on the journal. Use the accompanying sketch to make a *checking template* for checking journal radius. Make it from cardboard—the kind that comes inside shirts from the laundry or thin sheet metal if you want something more durable. Maximum specified main-journal radius is 0.100 inch. If any crank radius isn't right, return the crankshaft to be corrected, otherwise it could spell trouble.

Smooth and Clean Journals and Seal Surface?—A newly reground crankshaft should have its journals checked for roughness for no other reason than they may be *too rough*—just like you checked your old crankshaft, but not the same way. Rather than using your finger, rub the edge of a copper penny lengthwise on the journal. If it leaves a line of copper, the journal is too rough. If this is the case, consider smoothing the journals yourself rather than going through the

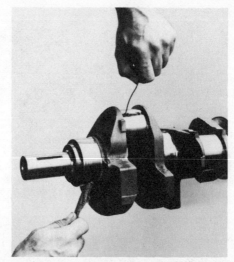
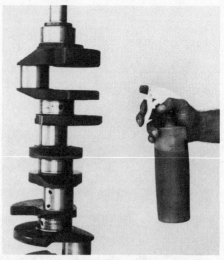

All that's normally required to restore a crankshaft: Lightly polish the bearing journals with fine-grit abrasive cloth. Clean the crank surface with special attention paid to the oil holes. A wire used to pull a solvent-soaked rag through the oil holes does a good cleaning job. A good coat of light oil keeps the journals rust free.

hassle of getting it redone. More often than not, you'll end up with a better job and avoid going through the frustration which usually accompanies getting the job redone. Also, if your original crankshaft checks out OK in all departments, you should give it the same treatment just to put a *tooth* on the highly polished journals and oil-seal surface, and to remove any varnish buildup.

Regardless of whether the journals are newly reground and need smoothing or are original, polish them with 400-grit emery cloth. A 1-inch-wide strip a couple of feet long should be sufficient to do all the journals and the rear-main seal surfaces. Wrap the cloth around the journal as far as possible and work it back and forth lightly as you gradually move around the journal. Keep track of where you start on a journal so you can give it an even finish all around. If you concentrate in one spot too long or use too much pressure, you'll remove material unevenly. The object isn't to remove material, but to give the journal and rear-main-seal surfaces a clean smooth surface. Be careful with the oil-seal surface. Don't do too good a job. Polish it just enough to remove any varnish or to smooth out any nicks or burrs. Some oil is required between the seal and the crankshaft to lubricate both. A highly polished surface will seal *too* completely with the result being eventual seal failure.

Crankshaft Runout—Crankshaft *runout* describes how much a crankshaft is *bent*. It is found by rotating the crank between two centers and reading runout with a dial indicator set 90° to the center main-bearing journal. As the crankshaft is rotated, the indicator reading will change if the crankshaft is bent.

I haven't mentioned runout until now because it's not likely you'll encounter this problem. If you do, it may be a false alarm because a cast-iron crank just lying around on the garage floor for some time can bend. Not enough so you can see it, but enough to show with a dial indicator.

Dial-Indicator Checking Method—All you'll need is an indicator with a tip extension so the crank throws won't interfere with the dial-indicator body as the crank is rotated. Just set the crank in the block with only the oiled *top halves of the front and rear bearing inserts* in place. Without the center three bearings in place, the crankshaft is free to *wobble* as you turn it. To measure this wobble or runout, mount your dial indicator base to the block with the indicator at 90° to the center main-bearing journal. Offset the indicator tip on the journal to miss the oil hole. Rotate the crankshaft until you find the lowest reading, then zero the indicator. You can turn the crankshaft and read runout directly. Turn the crank a few times to make sure you get a good reading. Maximum allowable runout is either 0.004 inch or 0.005 inch per the table on page 60. If your crank exceeds this, don't panic. As I said, a crank can change from just sitting around, so if yours is beyond the limit, turn the side that yielded maximum runout up, or down as the crankshaft would look with the engine in its normal position. Install the center main cap with its bearing insert. This will *pull* the crank into position. Leave it this way for a day or two and recheck the runout. Your crankshaft will probably *creep*, or bend to put it within the specified range. However, if it's too far off and can't be corrected using this method, have it reground or trade it in for a crank kit.

Cleaning and Inspecting the Crankshaft—When cleaning a crankshaft, it is very important to concentrate on the oil holes. Get them really clean. Even if you sent your crank along with the block for cleaning, some wire tied to a strip of lint-free cloth pulled through the oil holes will remove what's left if the cleaning solution didn't get it. Soak the rag in carburetor cleaner or lacquer thinner and run it through each hole several times. If you want to be sure the oil holes are really clean, a copper-bristle gun-bore brush works great. You may have used one of these for cleaning your block's oil galleries. By all means use it if you have one.

Installation Checking Method—To do a real-world check, install your crankshaft in the block using *oiled new bearings,* but *without the rear-main seal.* Torque the caps to specification. You'll find these specs in the assembly chapter, page 101. If the crank can be rotated freely by hand, consider it OK. Any loads induced by what runout there is will be minor compared to the inertial and power-producing loads normally applied to the journals and bearings when the engine is running. I suggest this method of checking because crankshaft runout isn't usually a problem with the normal "tired" engine that just needs rebuilding. If you decide to use this check, follow the procedure for crank installation detailed in the assembly chapter.

Using grease and some machined round stock to force the pilot bearing out of the back of this crank. If you use this method, just make sure you totally fill the crankshaft cavity and bearing bore with grease before starting to drive it out.

Using a socket and a brass mallet to install a new pilot bearing. Be careful not to damage the bearing bore when installing yours. Drive bearing in until it bottoms squarely in crankshaft.

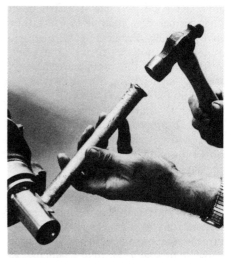
If you removed crank key/s for any reason—429/460 has two—reinstall it after checking for and removing any burrs in keyway slot. Tap lightly to get it started, then drive it in with something soft like a brass punch until it bottoms.

Pilot Bearing—With a standard transmission, a pilot bearing or bushing will be installed in the center of the flywheel flange. It supports the front of the transmission input shaft which pilots into it, thus the name *pilot bearing*. Visually check its bore for damage. If you have an old transmission input shaft, insert it into the bearing and check for lateral play by trying to move the shaft side-to-side. If there is noticeable movement, the bearing should be replaced. For a more accurate check, measure its bore with a telescoping gage and micrometer. The bearing's standard bore diameter is 0.625 inch and its maximum allowable diameter is 0.628 inch.

To be on the safe side I recommend you replace the pilot bearing and don't bother to check it. The cost is minimal and it's a lot easier to do now rather than after your engine is back in place and in operation. You can remove a pilot bearing by using any one of four different methods—there are more, but these are the most common. Two require special tools—a *puller* or a *slide hammer*. The slide hammer shaft screws into or hooks onto the back of the pilot bearing. A heavy sliding handle on the shaft bottoms against the end of the shaft to knock the bearing free. The puller screws into the pilot bearing and uses a steady force to pull the bearing. Tightening a nut on the puller gradually pulls the bearing out of its bore.

Two other methods employ the same basics. The first involves filling the pilot-bearing bore with grease. Then insert either an old input shaft, a short 5/8-inch diameter bar or a bolt (with its threads cut off) in the bearing bore and hit the bar or bolt with a hammer. Hydraulic pressure will force the bearing loose. The final method uses a coarse-thread 11/16-inch bolt ground with a long taper on its first few threads. Thread the bolt into the bearing until it bottoms on the crank. Keep turning the bolt until it pushes the bearing out. Regardless of the method you choose, note how deep the bearing was positioned in the crank and replace it in the same position. Normally, it will be flush with the end of the crank.

Prior to installing your new bearing, soak it in oil. The bearing is made from Oilite, a special bronze which is self-lubricating. Soaking it with oil gives it additional lubrication. Also grease the bearing bore in the end of the crankshaft to assist in installing the bearing. Line the bearing up with the crank bore and tap it into place. Use a brass mallet and a thick-wall tube or pipe which fits the outer edge of the bearing to prevent damaging the bearing's bore while installing it. A large socket works well for this if you don't mind pounding on it.

When starting to drive the bearing in, go easy at first. Be careful not to cock the bearing. Once it is started straight, it should go the rest of the way. If it cocks in the bore, stop and remove it, then start over. Drive the bearing in until it is positioned correctly.

CAMSHAFT & LIFTERS

After leaving a tough customer like the crankshaft, we now come to one not so tough—the camshaft and its lifters. As opposed to a crankshaft, once a cam starts wearing, it's not long before the cam lobes and lifters are *gone*. Consequently, you must give your old camshaft and lifters a very good look before deciding to reuse them. Two hard and fast rules apply to camshaft and lifter usage:

1. When reusing a camshaft and its lifters, **never install the lifters out of order.** The lifters must be installed on the same lobes, or in the same lifter bores.

2. **Never install used lifters with a new camshaft.** I also recommend not using a

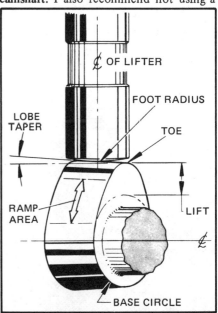
Visually inspecting camshaft-lobe wear pattern and lifter foot will tell you whether your cam should be replaced. If wear pattern is the width of the cam-lobe over most of toe area and lifter foot is worn flat or concave, your camshaft *and* lifters need replacing.

used, but good cam and its lifters in an engine other than the one in which they were originally installed. The lifter bore centers will not be in the same *exact* relationship to the cam lobes as in its original engine. So this is similar to mixing up the lifters.

Almost without exception, the rules realted to proper cam and lifter combinations and installation methods have to do with avoiding excessively high contact pressure between cam lobes and lifters. Loads exerted between a cam and its lifters during normal operating conditions are extremely high. Consequently, if there's any wrong move with the choice of cam components and their installation between now and the first 30 minutes of engine operation, the chances of a camshaft and its lifters being ruined are very high. It's important to adhere to proper procedures when choosing and installing your camshaft.

Camshaft Lobe and Lifter Design—For a clear understanding of why certain things have to be done when dealing with your engine's camshaft and lifters, a quick lesson in cam lobe and lifter design is in order. First is the *profile* of a lobe, or what one looks like when viewed from the cam end. A cam's profile governs *how far* its valves open in addition to *when* they open and close. This is an oversimplification, but it's OK for our purposes.

When a valve is closed, the lifter is on the *base circle* of the cam lobe described by a radius. When the lifter is on the highest point of the lobe, it's on the *toe* of the lobe and the valve is fully open. Between the base circle and the toe are the opening and closing *ramps*. Sketch on page 63 shows the difference between the *base-circle diameter* and the distance from the toe directly across the camshaft centerline to the base circle is cam lift at the lifter—*not the valve*.

During the camshaft manufacturing process, lobe surfaces are not ground parallel to the camshaft centerline, but are ground at a *rake angle*—1° in most instances. Also, the lifter *foot* that contacts the cam is not ground flat, but has a *spherical radius*, or is ground *convex*. Ford uses a 30-inch radius. At first this sounds odd because one would think pressure between the lobe and lifter would go up as area between the two is reduced. However, this machining ensures a reasonable and predictable contact pressure within practical machining tolerances. The lifter-foot radius and lobe angle guarantee good cam-to-lifter contact area for consistent camshaft and lifter life.

The way camshafts and lifters are machined serves another important function. Cam-lobe contact is made *off-center* rather than directly on the lobe's center. Consequently, the lifter rotates to minimize wear of the two components and to ensure good lifter and bore lubrication.

CAM & LIFTER INSPECTION

During their operating life, a cam and its lifters wear gradually. How much depends on how an engine is operated and, more importantly, how it is maintained. You must determine whether yours has worn so badly that it shouldn't be used.

I'll say it right now. Using an old camshaft and lifters in a newly rebuilt engine is a risky deal. It's not uncommon for a new cam to fail during its first 100 miles—or even before it gets out of the driveway. Finally, if you've lost track of the order of your lifters, toss them away and get new ones. I recommend going the full route and getting a new cam too because the lifters are the expensive part. The odds of getting 16 lifters back in the right order on 16 cam lobes are absolutely astronomical: 20,922,789,980 to 1.

Check the Camshaft First—When checking your cam and lifters, check the cam first because it it's bad you'll have to replace the cam *and the lifters*, regardless of lifter condition. Remember the second rule: *never install used lifters on a new cam.*

The first thing you do is get out your trusty micrometers—vernier calipers are OK—and check your cam's lobe lift. Maybe you did this during the diagnosis process prior to removing and tearing down your engine. You'll have used a dial indicator at the end of a pushrod to check the lobe lift the rocker arm sees, or *actual* lobe lift. If you used this method and made a determination about your cam, you can forget measuring the lobes directly because lift measured at the lifter or pushrod is accurate. On the other hand, using verniers or mikes becomes increasingly inaccurate with high-lift, high-performance camshafts because the ramp area extends farther around the lobe and reduces the length of the base circle.

After arriving at a figure for each lobe, compare all the exhaust-lobe figures. If the numbers aren't within 0.005 inch you know some lobes are worn excessively and the cam should be replaced. Lobe positions for intake and exhaust valves, from either end of the camshaft are: I E E I I E E I I E E I I E E I.

Another thing to note is the wear pattern on each lobe, particularly at the toe. If the wear has extended from one side of the toe to the other and well down onto the ramp area, the cam has seen its better days and should be replaced, even though its lift checks OK. This goes for pitting too. *Any* pitting on a lobe's *lift areas* indicates metal loss which will probably show up on its mating lifter. A "full-width" wear pattern or pitting of a cam's lobes indicates the cam and lifters should be replaced.

Cam-bearing journals are "bullet proof"—they never seem to wear out—at least I've never seen any worn out. However, if you feel compelled to check yours

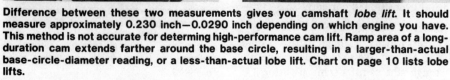
Difference between these two measurements gives you camshaft *lobe lift*. It should measure approximately 0.230 inch—0.0290 inch depending on which engine you have. This method is not accurate for determining high-performance cam lift. Ramp area of a long-duration cam extends farther around the base circle, resulting in a larger-than-actual base-circle-diameter reading, or a less-than-actual lobe lift. Chart on page 10 lists lobe lifts.

while your micrometer is handy, do it. Checking never hurt anything and the key to building a good engine is check, check and recheck. Just like the crankshaft, there are five camshaft bearing journals. Journal diameters for 429/460 camshafts are all the same size at 2.1238–2.1248 inches. 351C, 351M and 400 cam journals are all different sizes with these limits:

BEARING JOURNAL	JOURNAL DIAMETER (inches)
1 (front)	2.1238–2.1248
2	2.0655–2.0665
3	2.0505–2.0515
4	2.0355–2.0365
5	2.0205–2.0215

Cam journals also have maximum out-of-round and runout limit specs like the crank: 0.0005-inch out-of-round and 0.005-inch runout. Again, it's unlikely these limits will be exceeded unless your engine has been "gone through" previously and a many-time-recycled cam installed. If this is the case, you'll be better off replacing the cam anyway. When checking cam-journal out-of-round, use the same method used for checking crankshaft journals—find maximum and minimum dimensions and subtract to get the measurement.

As for runout, you can't check it unless you have centers to mount and rotate the cam on—a lathe works fine for this. A surface plate, angle blocks and a dial indicator can also be used. However, don't worry about it. Excessive cam runout is the least of your worries because it seldom occurs. The real check is whether the cam turns easily when installed in the block. If it does, consider it OK.

Check Your Lifters if the Cam is OK—I may sound like a broken record, but be triple-careful not to get the lifters out of order. If you do, you must pay the price and replace them, even if the cam checks OK. Of course, you won't have to bother checking the lifters if your cam is not reusable. However, if you can reuse the cam, proceed.

The thing you're looking for when checking a lifter is a spherical radius on the foot. If the radius is gone, the lifter is junk. Because this radius is so large—about 30 inches—it's difficult to check with a straight edge, even if the lifter is brand-new. So double the effective radius by butting two lifters together end-to-end. They will rock back-and-forth if any radius exists. If they won't rock, the lifters are either worn flat or concave and one or both of the lifters is junk.

I hate to keep repeating myself, but because of the touchy nature of a camshaft you must be aware of the consequences. If a lifter is worn concave, the lifter must be replaced, and the cam also—even though lift checks OK. If this is the case, the toe of the lobe this lifter mated with will be worn all the way across.

Clean the Lifters—Concern yourself with cleaning your lifters only if your cam is reusable and the lifters are OK. Clean the varnish buildup from around the foot of each lifter and any sludge or varnish which has accumulated inside the lifters if they are the hydraulic type. The outsides will come clean by soaking them in carburetor cleaner.

Soaking doesn't get the inside at all. You can take two approaches here. The easiest way to clean a hydraulic lifter is to use an oil squirt can to force lacquer thinner through the hole on the side of the lifter body as you work the plunger up and down until it feels free. The most positive lifter-cleaning method is to take the lifter apart, but it's also difficult. There are eight parts to contend with, starting with the retaining clip inside the groove at the top of the lifter body. Be careful during disassembly so you don't damage any of the lifter components or lose any of the small parts. Disassemble and clean one lifter at a time so there's no chance of mixing up the parts because they are matched sets. You'll need small needle-nose pliers and a lot of patience, so much that you may decide to revert to the lacquer thinner and squirt-can approach. If you do decide to continue with this method, you may have to use the squirt can to help loosen the plunger so it will come out of the lifter body. Do this after you've removed the retaining clip, pushrod cup and metering valve. By working the plunger up and down with a few injections of thinner into the body, the plunger should gradually work its way out. Besides the plunger, you'll have more parts to contend with as the plunger comes out. They are small and easy to lose, so be careful.

After cleaning each lifter component separately, reassemble that lifter *before* starting on another one. This prevents mixing components and it's easier to keep the lifters in order too. To assemble a lifter, stand the plunger on your work bench upside down and assemble the parts that go below it in the lifter body onto the plunger. Depending on the type lifter you have, there will be a check valve, a small check-valve spring, and/or a check-valve retainer and a plunger spring—in that order. With these stacked on the plunger, oil the lifter-body ID and slide it down over the plunger, check valve and spring assembly. Turn this assembly over and install the metering valve and disc or just a metering valve, depending on the type lifter you have, the pushrod cup and retaining clip. One lifter cleaned and assembled. Do this 15 times more to finish cleaning your lifters.

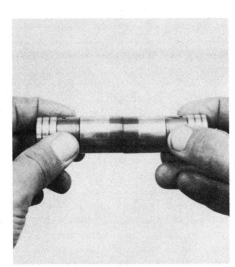

Putting foot-ends of two old lifters end-to-end tells you if any spherical radius is left. If you can't rock them relative to one another, they shouldn't be reused and your lifters *and* cam must be replaced.

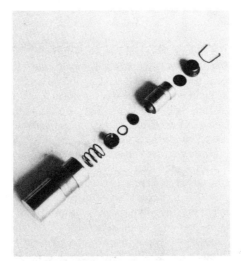

Parts of a hydraulic valve lifter. If you disassemble yours for cleaning, make certain you don't lose any of the parts or interchange plungers between lifter bodies.

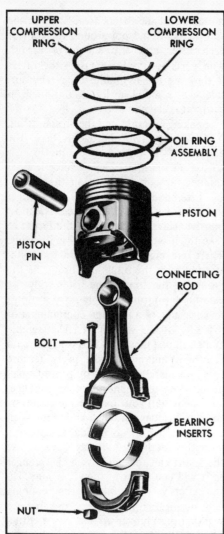

Complete piston and connecting-rod assembly including bearing inserts and piston rings. Wrist pin is prevented from moving laterally by an interference fit between it and its connecting-rod bore. Photo courtesy Ford.

PISTONS & CONNECTING RODS

Replace the Pistons?—If your block has to be rebored for one reason or another, you'll have to replace your pistons to fit the larger holes. However, if you determine from checking that most of your pistons should be replaced, then the block should have been bored in the first place. The cost of pistons is the major expense, so you may as well give your engine a fresh start with new pistons and *straight* bores. Also, its durability will be as good or better than new, depending on the care you take during the rebuild.

CHECKING THE PISTONS

If your engine doesn't need reboring, your next step is to check the pistons to establish whether or not they are reusable *before disassembling them from the rods.* Remove the old rings and toss them away. Be careful when doing this so you don't scratch the pistons in the process. A *ring expander* will help. Don't remove the rings down over the skirt, remove the top ring first, then the second followed by the oil ring. If you do this by hand, make sure the ends of the rings don't gouge the piston. For convenience, support the piston-and-rod assembly so it doesn't flip-flop while you're trying to remove the rings. Clamp the rod *lightly* in a vise with the bottom of the piston against the vise. If you don't have a vise, clamp the rod to the edge of your workbench with a C-clamp—again not too tightly.

Four items should be checked before a piston is given the OK: general damage to the dome, skirt or ring lands; ring-groove wear; piston-skirt and pin-bore wear. If any one proves unsatisfactory, replace the piston.

General Damage—Not so obvious damage that can render a piston useless: skirt scuffing or scoring, skirt collapse, ring-land damage and dome burning. Obvious damage can be done by something such as a valve dropping into the cylinder.

Scuffing and scoring is caused by lack of lubrication, excessively high operating temperatures or a bent connecting rod. All cause high pressure or temperature between the piston and cylinder wall. If there are visible scuffing or scoring marks—linear marks in the direction of piston travel—replace the piston. Scuff-marking indicates the engine was excessively overheated. If the piston has *unsymmetrical* worn surfaces on the skirt thrust faces, a twisted or bent connecting rod is the likely culprit and should be checked and corrected. An engine machine shop has the equipment to do this and it should be a normal part of their routine when rebuilding an engine to check all the connecting rods for alignment and bearing bore, regardless of what the old pistons look like. However, if you are on a tight budget, the wear pattern on the pistons will tell you what you want to know. Otherwise, it's not a bad idea to have your rods checked.

Damage to a piston's dome usually comes in the form of material being removed as a result of being overheated from detonation or preignition. The edges of the dome will be rounded off or *porous* or *spongy-looking* areas will show high heat concentration. To get a good look at the dome, clean off any carbon deposits. A good tool is a worn screwdriver which has rounded corners at its tip. The normal toolbox is usually well equipped with these. Be careful when

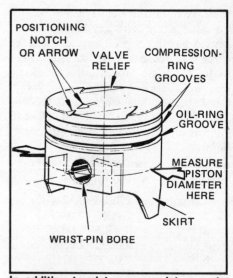

In addition to piston nomenclature, note positioning notch or arrow in top of piston. It is important for assembling a piston to its connecting rod, then for installing complete assembly in its bore.

scraping the carbon so you don't damage the piston by digging into the aluminum. Also, don't use a sharp or hard tool like a chisel or gasket scraper for this very reason—or be extra careful.

Detonation or explosion of the fuel charge, can also cause broken or distorted ring lands through impact loading. Check the top ring land for this condition. It receives the brunt of the compression loading. Consequently, if it's not damaged, the others will be OK. Broken ring lands are readily visible, but a bent one may not be, particularly without a ring in the groove to use as a reference. Reinstall a ring in the top groove and use a feeler gage which fits snuggly between the ring and the groove. Slide the ring and feeler gage around the groove to check for any ring side-clearance changes which may indicate a distorted ring land. The top ring land is also the one that gets the wear from an engine inhaling dirt. The wear will be on its upper surface and even all the way around.

Any of the types of damage I have just mentioned are causes for discarding a piston.

Piston-Skirt Diameter—It's micrometer time again. Mike each piston 90° to its connecting-rod pin axis in the plane of the pin and compare this figure to what the piston mikes across the bottom of its skirt. If the skirt is not wider by at least 0.0005 inch than at the pin, the piston should be replaced because this indicates the skirt is partially collapsed. Skirt collapse is usually accompanied by heavy scoring or scuff marks on the skirt—sure signs that the engine was severely over-

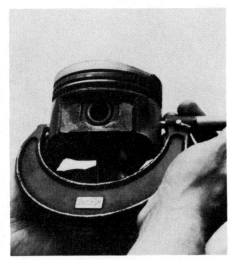

When measuring piston diameter, do it level with the wrist-pin and 90° to pin axis. Width measured across skirt bottom should be slightly larger—about 0.0005 inch.

Scuff marks on piston skirt reveal engine was severely overheated. Further use will overstress and possibly collapse skirt. Scrap any piston that looks like this.

Wear pattern on a piston skirt indicates connecting-rod alignment. If the pattern is symmetrical about the centerline of the skirt, rod alignment can be considered OK. However, a skewed pattern like this shows the rod is bent or twisted and should be straightened. A checking fixture is required for this.

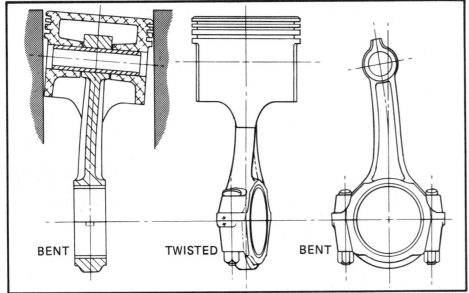

First two rods, bent and twisted as shown in these exaggerated drawings, cause uneven piston-skirt wear by tilting piston in bore. First rod, with its wrist-pin axis bent out of parallel with its bearing journal, tilts its piston mostly at TDC and BDC. Twisted connecting rod tilts progressively toward the mid-point of stroke. Connecting rod bent like third one shown doesn't affect skirt wear or engine performance. Consequently it can be reinstalled unless the bend is obvious to the eye, or unless it's to be used in an all-out high-performance application.

heated at least once. If a piston with these symptons were to be reinstalled, you'd have a very noisy engine which may eventually experience total piston-skirt failure.

Piston scuffing or scoring occurs when an engine overheats. The piston tries to expand more than its bore. If the engine is heated too much, the piston skirt squeezes out the oil cushion between the piston and the bore. Not only does the skirt make metal-to-metal contact with the bore, it is overstressed so both scuffing or scoring and skirt collapse occur. After the piston returns to its normal temperature, its bore clearance will be excessive and the engine will be noisy, particularly during warmup.

Piston-Pin Bore Wear—Measuring pin-bore wear requires disassembling the piston and rod. A practical and easier way of doing this is to *feel* for wear. The normal pin clearance is 0.0003 inch and should not exceed 0.0008 inch. Excess clearance will show up when you try wiggling or rotating the connecting rod 90° to its normal direction of rotation. First, clean each piston with solvent to remove any oil between the piston and the pin. Oil takes up clearance and sufficiently cushions movement to give the impression the piston-to-rod relationship is OK, even when it's not. Now, with the piston and rod at room temperature, if you feel movement as you try wiggling the rod sideways, the pin bore is worn beyond the limit and the piston should be replaced. Otherwise it's OK.

After you use this method to check for pin-bore wear, make sure you lubricate the pin and its bore by squirting oil into the oil groove in the piston-pin bore while working the rod back-and-forth. If you don't you'll run the risk of ruining some pistons during initial engine startup. The pin and its bore are highly loaded. Consequently, if they are not lubricated when you first fire your engine, the aluminum will gall causing the piston to seize on the pin. Lubricate the pins now so you don't overlook it later on.

Clean the Ring Grooves First—I saved this job till last because cleaning the ring grooves is a tough job and you have to do it to do a complete job of checking the grooves. So, if some or all of your pistons didn't pass your previous tests, you've avoided unnecessary work.

You'll need something to clean the ring grooves without damaging them. A special tool for doing just this is appropriately called a *ring-groove cleaner*. Prices range from $10 to $30, depending on tool quality. A ring-groove cleaner fits around the piston and pilots in the groove it is cleaning. An adjustable scraper fits in the groove and cleans carbon and sludge deposits from the groove as the cleaner is rotated around the piston. If it bothers you to make a purchase like this for a "one-time" use, use the broken end of a piston ring to clean the grooves. It takes more time and you'll have to be careful, but it can be done just as well. Grind or

BLOCK RECONDITIONING 67

file the end of the ring like the one shown in the accompanying photo. Wrap tape around the ring so you don't cut your fingers on it.

Be careful, no matter which method you use. Caution: Don't remove any metal or scratch the grooves, just get rid of the deposits. Be especially careful to avoid removing metal from the side surfaces of the grooves as they are the surfaces against which the rings seal.

Measuring Ring-Groove Width—Again, chances are if a ring groove is worn or damaged, it will be the top one because of its higher loads. So start with it. This doesn't mean you don't have to check the others if the top one is OK. Unforeseen things can happen, so check them all. Look at the clean grooves first. Any wear will have formed a *step* on the lower portion of the ring land. The height of the step shows up as additional ring *side clearance* and the length of the step projecting from the back-wall of the groove represents the piston ring's back-clearance. Side clearance between one side of the ring and a ring land is measured with a ring and feeler gage with the ring held against the opposite ring land. If it is too much, proper ring-to-piston sealing is not possible, plus the additional action of the ring moving up and down in the groove will accelerate wear and increase the possibility of breaking a ring land.

Compression-Ring Grooves—Compression-ring grooves are nominally 0.080-inch wide, new rings have a minimum 0.077-inch width and a maximum ring side clearance of 0.006 inch. You are presented with a dilemma because the 0.006-inch side clearance should be measured with a new ring in the groove. You don't want to lay out money for a new set of rings for checking your old pistons until you've checked them and have given them the OK. One way around this is to use an old ring for checking. The problem is, rings wear too, and more than actual side clearance will show up if ring wear is not accounted for. Just mike the old ring and *subtract* this amount from 0.077 inch, the minimum width of a new compression ring. Add this figure to 0.006 inch for the maximum allowable *checking clearance*. I'll use the term *checking clearance* rather than *side clearance* because the gage thickness may not be the actual side clearance with a new ring. If the old ring measured 0.076 inch, it is 0.001-inch undersize. Consequently, maximum checking clearance is now 0.007 inch but it will yield a 0.006-inch side clearance with a *minimum-width*

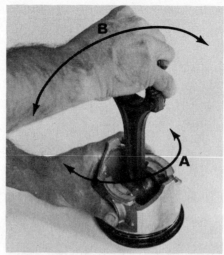

Checking pin-bore wear by twisting connecting rod in direction A, then trying to rotate it in direction B, 90° to its normal direction of rotation. If you can feel movement at pin when holding piston like this, pin-bore wear is excessive and the piston should be replaced.

Start cleaning your pistons by removing carbon from their tops. A dull screwdriver is good for removing big chunks and a wire brush does cleanup work, but don't touch piston sides with the brush. A piston can easily be ruined by careless wire brushing.

Two tools which accomplish the same thing—an official ring-groove cleaner or a broken piston ring ground like this. Ring is harder to use, but a lot cheaper. Regardless of how you clean your ring grooves, be careful not to remove any metal.

new ring. In formula form, this looks like:

Maximum checking clearance
= 0.006 + (0.077 − 0.076) or:
Maximum checking clearance in inches
= 0.006 + (0.077 − checking-ring width in inches)

When checking side clearance, insert the edge of the ring in the groove with the checking-clearance feeler gage alongside *between the ring and the lower side of the ring groove*. Slide the gage and ring completely around the groove to check for any clearance variations and to verify the land is free from distortion or uneven wear. The gage should slide the full circumference without binding. Check both compression-ring grooves using this method. Make sure the ring is not up on the step when checking side clearance.

Oil-Ring Grooves—Nominal oil-ring grooves are 0.1885-inch wide. The real test is a snug fitting oil ring. Here's that old problem again. You don't yet have a new set of rings to do the checking. Fortunately oil rings and their grooves are well lubricated and aren't heavily loaded like the compression rings, particularly the top one. Consequently, their wear is minimal, so if they pass your visual inspection you can assume they are all right. However, to be positive, measure them. To do this, stack two old compression rings together and mike their combined thickness.

Using a piston ring and a feeler gage for checking ring-groove wear. New rings should have no more than 0.006-inch side clearance.

Using a connecting-rod alignment-checking fixture to check for bent or twisted connecting rods. When using this type of checking fixture, the piston must be installed on the rod. If a rod is found to be out of alignment, it is straightened and then rechecked.

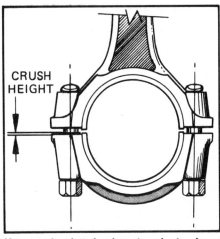
How much a bearing insert projects above its bearing-bore parting line is called *crush height* (about 0.001 inch). As bearing cap is tightened down, the insert-half ends contact first, forcing inserts to conform to shape of bearing bore and to be locked in place. This is called *bearing crush*.

As an example, if the combined thickness of two rings is 0.155 inch, and feeler-gage thickness to get maximum oil-ring-groove width, or 0.191 inch minus 0.155 inch = maximum allowable checking feeler-gage thickness of 0.036 inch.

Maximum checking gage
thickness in inches
= 0.191 − Combined thickness of
two compression rings in inches

If the oil-ring grooves in your pistons don't exceed this amount, you can be certain the rings will fit snugly in their grooves.

Now, use stacked rings with your feeler gage to check groove width. You don't have to install the rings in the groove. Just inset their edges in the groove with the gage and slide them around in the groove as you did when checking the compression-ring grooves.

Groove Inserts—If you have the unlikely circumstance that your cylinders don't need boring, only honing to get them back into condition, and your pistons are OK for reuse except for having too much ring-groove wear, you can have your piston-ring grooves machined wider. This makes all the ring-groove widths consistent—and larger. To compensate for the additional clearance between the rings and the grooves, ring-groove inserts are installed—they are usually 0.060-inch wide and are installed beside the rings. In terms of cost, ridge-reaming and honing your block, machining your pistons in preparation for ring-groove inserts and purchasing the rings and inserts will cost 50—60 percent of a rebore and new pistons and rings. In this case, durability is directly related to cost. You can expect approximately half the life from an engine with inserted pistons compared to one with a complete rebore job.

CONNECTING RODS

Inspecting connecting rods involves checking three areas: out-of-round or enlarged bearing bores, twisted or bent rod beams and cracked rod bolts. Now's when the old bearing inserts come in handy. If a rod bearing shows uneven wear from side-to-side—opposite sides on top and bottom bearing halves—the piston on that rod has wear spots offset from its thrust face and the rod's crankshaft bearing journal was not *tapered*, the rod is bent, causing *side loading*. With these symptoms, the rod and piston assembly or assemblies should be taken to an engine machine shop for accurate checking and straightening if necessary. Bearing condition also tells you if the "big end" of the rod needs *reconditioning*. When a rod is reconditioned the bearing bore, or big end is checked with a special dial indicator to determine its shape—round, out-of-round or oversize. This is basically the same thing you did with the crankshaft bearing journals, however now the check is made of the bore in which the bearings are retained. With an out-of-round bore, its bearing inserts will assume the same irregularity, causing uneven load distribution between the bearing and its journal, resulting in uneven and accelerated bearing wear. Consequently, the bore must be reconditioned.

A bearing with a too-large bore is even worse. Oversize bearing bores will let bearings move in their bore—which is not supposed to happen. The reason is the insert halves are not sufficiently *crushed*. Crush occurs when a bearing cap is torqued, *forcing* the bearing to conform to its bore. This is accomplished by the combined outside diameter of the two bearing insert halves being larger than the ID of the bore. When a bearing half is placed in its bore, its ends project slightly above the bearing housing's parting surfaces. Consequently, when two bearing inserts are installed in their bore, the ends of the inserts butt. As the bearing-cap nuts are tightened, the two circumferences must become equal. The bearing shell gives—is crushed—causing the bearing to assume the shape of its bore and to be preloaded, or fitted tightly in its bore. This tight fit and the tooth, or machining marks in the bearing bore, combine to prevent the bearing inserts from *spinning*—rotating in the bore rather than the bearing journal rotating in the bearing. This happens when the force at the bearing journal which tries to rotate the bearing overcomes the force between the bearing and its housing, or bore that is resisting this force. Consequently, if the bearing bore is too large, the bearing may move in the bore, or worse yet, spin.

To determine if a bearing has been moving in its bore, look at its backside. Shiny spots on the back of the shell indicate movement. If this happened, either the bearing bore is too large or the bearing-to-journal clearance was insufficient. If you discover any shiny spots, the rod/s need to be checked with a dial-indicator that's part of an engine builder's connecting-rod conditioning hone. A telescoping gage and micrometer can also be used for checking. Recondition the rod/s as necessary.

When a connecting rod is reconditioned, some material is precision ground from the bearing-cap mating surfaces. Next, the cap is reinstalled on the rod, bolts are torqued to spec and the bore is honed to the specified diameter. Honing corrects the bearing-bore diameter and concentricity and also restores the *tooth*, or surface of the bore which *grabs* the bearing insert to prevent it from spinning.

Rod Bolts—Make sure you inspect the rod bolts very closely *before* attempting any other reconditioning of the rods. Don't hesitate to use a magnifying glass as you give them the old "eagle-eye." Replace any cracked bolts—a cracked rod bolt will eventually break and may totally destroy any engine.

You have to remove the bolts to check them. To remove a rod bolt, clamp the big end of the rod in a vise between two blocks of wood and drive the bolt straight out. Use something such as a brass punch to do this to prevent damaging the bolt. The original or new bolt can be installed by tightening it after it is loosely installed in the rod with the cap. If you replace one or both rod bolts, the connecting rod *must be reconditioned* or honed because the bolts locate the cap in relation to the rod. Consequently, changing a bolt may shift the cap in relation to the connecting rod.

One last note concerning connecting rods. Rod-bearing bores are originally machined with their caps torqued in place, so they must be checked the same way. When you deliver them to a machine shop for checking and reconditioning after you've inspected the rod bolts, make sure the correct cap is installed in each rod. Torque the nuts to specification before making the delivery; 40–45 ft. lbs.; 43–48 ft. lbs. for 351C Boss and HO.

DISASSEMBLING & ASSEMBLING PISTONS & CONNECTING RODS

Disassembling or assembling connecting rods and pistons is a job for an expert with the proper equipment. If you are replacing your pistons, the rods and pistons must part company. Only one method can be used to do this correctly. The connecting-rod wrist pin must be pressed—**not driven**—out of the small end of the rod in which it is retained by 0.001–0.0015-inch interference fit. A

 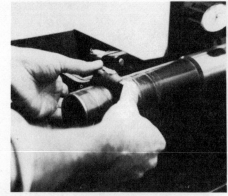

Connecting-rod bearing bore being checked and honed, or reconditioned. If bearing bore isn't nearly perfectly round, some material is precision ground from bearing-cap parting surfaces. Cap is then reassembled to rod and nuts torqued to spec. Bearing bore is then honed to its correct diameter. When reusing old pistons, they need not be removed from their connecting rods for reconditioning the rods.

If your pistons have to be replaced, this is the first operation you'll have to do—disassemble them. A press and mandrel are required. You can't drive the wrist pins out with a hammer and punch, so don't even try.

CONNECTING-ROD BORES DIAMETERS (INCH)

	Wrist Pin	*Bearing
351C	0.9104–0.9112	2.4361–2.4369
351M/400	0.9726–0.9742	2.4361–2.4369
429/460	1.0386–1.0393	2.6522–2.6530

*Maximum out-of-round 0.0004 inch.

If the difference between the high and low bearing-bore measurements at a connecting rod's "big end" exceeds 0.0004 inch, the rod needs to be reconditioned. The same applies if the bore is not within the diameter specified.

Notch or arrow, indicates correct position of a piston in its bore. Notch or arrow should point to engine front. When assembling connecting rods and pistons, rod numbers 1, 2, 3 and 4 must be to the right of the piston as you view it holding piston so notch or arrow is pointing away from you, and numbers 5, 6, 7 and 8 should be to the left. As you can see from the box label and number stamped in the piston top, this one is for a 0.030-inch oversize bore.

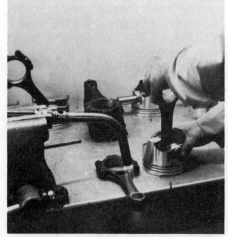

Piston and connecting rod being assembled by enlarging the wrist-pin bores by heating with a propane torch. Pin slips right into place, but you've got to be fast. Rod and wrist-pin temperatures converge very quickly, causing the two to lock together within a couple of seconds. If pin is not located when this happens, you'll then have to resort to a press and mandrel.

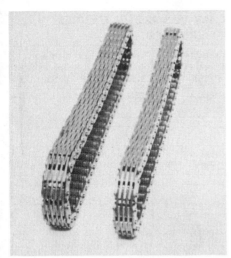

Five- and four-plate timing chains for use with 1/2-inch and 13/32-inch-wide sprockets. Narrow chain and sprockets were used only in '73 and later 351C, 351M and 400 passenger-car engines.

press with mandrels to back up the piston and bear on the pin is required.

Assembling a connecting rod and piston can be accomplished by one of two methods. The first is done by reversing the disassembly process, however it must be done with considerably more care. The other method, and the one I prefer, is the heating method. This is done by simply heating the small end of the connecting rod with a torch so it expands, allowing the pin to slide into place without need for a press. The rod should not be heated excessively because of the possible damage. Pin and piston have to be fitted very quickly to the rod before the rod and pin approach the same temperature. Otherwise the pin is trapped by the interference fit before it's in position, requiring a press to complete the installation.

If you are assembling your own pistons and rods, be aware of the piston's location relative to its connecting rod. As installed in the engine, pistons usually have a notch at their front edge or an arrow stamped in the dome, both of which *must point to the front of the engine* as installed. If you have the Boss 351C, or the 429 CJ, SCJ or Police engines, the single intake-valve reliefs should be on the intake-manifold side of the bores. Also, connecting-rod numbers *must point toward their cylinder bank.* Numbers on rods installed in the right bank must point to the right and those in the left bank must point to the left. So, putting them together, you'll have piston and rod number 1, 2, 3 and 4 with *notches or arrows pointing forward and rod numbers to the right.* Piston and rod numbers 5, 6, 7 and 8 have *piston notches or arrows pointing forward, and rod numbers to the left.* As for the Boss 351C, 429 CJ, SCJ and Police, the valve reliefs should be located opposite the rod numbers.

TIMING CHAIN & SPROCKETS

I previously discussed timing-chain wear and how to determine if a chain and its sprockets need replacing. However, here's some more information about what replacement parts you should consider. If your engine is one of the hotter operating '73 or later engines and you live in a hot climate, change to the cast-iron cam sprocket. This avoids the problem of cracked and broken sprocket teeth, but don't expect better sprocket or chain wear. There's little if any difference.

Several changes have been made to the 351C, 351M/400 and 429/460 cam-drive components over the years. Some changes have been major to the eye, with little effect on the engine performance. Some subtle changes have affected engine performance considerably. For instance, the '73 351C-2V went to narrow 13/32-inch-wide sprockets and a 4-plate-wide chain as compared to the 1/2-inch sprockets and 5-plate chain previously used. This change was also carried over into the 400 and was originally installed on the 351M. For a clearer picture of the changes made in the timing-chain department, refer to the nearby chart because it does get complicated. For instance, 351Ms and 400s installed in trucks and Broncos have wider chains and sprockets while passenger-car engines have had their cam timing altered by changing the keyway position in the crankshaft sprocket. This is the subtle change I was speaking of. These types of changes were also made to the 429/460 engines except these engines stayed with the 1/2-inch-wide sprockets and chains. One major difference in 460s built for use in motor-home chassis — M450s and M500s—between 1974 and 1976 is they used a double roller-type chain and sprockets. This type chain is very durable and is normally available as an aftermarket item.

Generally, the changes made to crank sprockets only resulted in the camshaft being retarded for emissions reasons. Unfortunately, this type of change also hurts engine performance. This is illustrated by the change made to the 460 Police-interceptor engine in 1977. The C8SZ-6306-A crank sprocket, replaced by the retarded D2VY-6303-A sprocket in 1975, was reinstated in 1977 to improve engine performance. Passenger-car and truck 429/460 engines retain the D2VY-6306-A sprocket with exception of motor-home engines. They use roller-style chains.

Again, another thing to consider when replacing your cam sprocket is this is an opportunity to change to the cast-iron type, particularly if you live in a hot climate. TRW offers both nylon and cast-iron cam sprockets. This makes it convenient because the gear specified for a specific engine can be installed on that engine without any other hardware changes. In other words, chains and gears are manufactured to accommodate the hardware originally installed on the engines for which they are specified.

Heavy-Duty or Performance Applications—In addition to the OEM and OEM-replacement timing sets, high-performance and heavy-duty sprockets and chains are available. Consider these if your engine is going to be used for extremely hard service. They elongate or *stretch* less and are relatively unaffected by high engine-operating temperatures. The chain I am referring to is a double-roller type with matching sprockets, similar in design to bicycle chains except with two rows of rollers. These are manufactured by Cloyes Gear and Products, Inc., 4520 Beidler Road, Willoughby, Ohio 44094 and TRW, 8001 E. Pleasant Valley Road, Cleveland, Ohio 44131. These products should be available through your local engine parts supplier.

Crankshaft End-Play—I could wait until the engine buildup chapter to talk about camshaft end-play, but it's a good idea to find out if you need a part now instead of waiting for a long holiday or late Saturday night when everything is closed and you are in the middle of assembling your engine.

A camshaft doesn't have to be installed in an engine to check its end-play. Just install the sprocket on the cam with the thrust plate in its normal position. Now, use your feeler gages to check the clearance between the thrust plate and the front cam-bearing-journal thrust face. The maximum-thickness feeler gage represents camshaft end-play. End-play should be in the 0.001–0.006-inch range with a maximum of 0.009 inch.

OIL PUMP & DRIVE SHAFT

Lubrication is a major key to an engine's durability, and the oil pump is the heart of the lubrication system. Don't take any short cuts here. Check your oil pump for any internal damage such as grooving or scoring of the rotors or the

With early- and late-model 429/460 crankshaft timing-chain sprockets lined up, you can see the keyway is moved to retard the late-model camshaft 8°. Effect of retarding a camshaft is to move engine's "power band" higher on the RPM scale, leaving it with less "low-end" power.

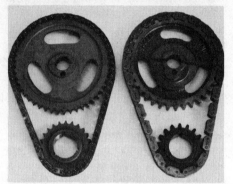

A high-performance roller-style timing-chain set compared to a conventional silent chain and its sprockets. If your engine will be subjected to severe service, consider using a roller-chain setup. Notice three keyways in roller-chain sprocket. One is for standard camshaft timing whereas others advance or retard cam, depending on whether engine's power is desired at lower or higher RPM, respectively. Roller timing-chain sets are available from Cloyes and TRW.

CAM DRIVE COMPONENTS 351C, 351M/400 & 429/460 ENGINES
(SERVICE PART NUMBERS USED)

Engine Year	Camshaft Sprocket (6256)	Chain (6228)	Crankshaft Sprocket (6306)	Thrust Plate 6269	Cam-Spr. Spacer 6265
351C2V *70/72	*D0AZ-A	D0AZ-A	C8SZ-A	C8SZ-A	Not. Req'd
351C2V •73/	•D3AZ-A	D3AZ-A	D5AZ-A	———	———
351C4V *70/74	D0AZ-A	D0AZ-A (36th)	C8SZ-A (48 Links)	(18th)	———
351 Boss *71/72	D0AZ-A	———	———	———	———
H.O. 351CJ *70	D0AZ-A	———	———	———	———
EX Bronco •351M 73/	D3AZ-A	D3AZ-A	D5AZ-A	———	———
Bronco *351M 78/	D0AZ-A	D04Z-A	D7TZ-A	———	———
Bronco *400	D0AZ-A	D0AZ-A	C8SZ-A	———	———
X Bronco •400 73/	D3AZ-A	D3AZ-A	BEF 8-1-75 D3AZ-A FROM 8-1-75 D5AZ-A	———	———
Bronco *400 78/	D0AZ-A	D0AZ-A	D7TZ-A	———	———
Truck *351M/400 77/	D0AZ-A	D0AZ-A	D7TZ-A	———	———
*429 68/78	D0AZ-Z	C8SZ-A (52 Links)	63/72 C8SZ-A 73/74 D2VY-A	———	C8SZ-A
*429CJ 68/73	———	———	———	———	———
*429SCJ 70/71	———	———	———	———	———
429 P/C 72	———	———	———	———	———
*460 68/72	D0AZ-A	———	68/72 C8SZ-A 73/78 D2VY-A	———	———
*460 P/C 73/78	D0AZ-A	———	73/74 C8SZ-A 75/76 D2VY-A 77/78 C8SZ-A	———	———
460 Truck *E250/350 F100/350 73/	D0AZ-A	———	73 C8SZ-A 74 D2VY-A (18th)	———	———
460 Truck Motor Home Chassis M450/500 74/76	D4TZ-A	D4TZ-A Roller-Style Chain	D4TZ-A (27th)	———	———

* 1/2" wide sprocket (5 plates)
• 13/32" wide sprocket (4 plates)

Use this chart to match camshaft-drive components. Numbers shown are current Ford service-part replacement numbers. If you order a complete timing-chain-and-sprocket set, you shouldn't have any matchup problems.

housing, and check clearances before reusing it. While it's one of the most durable components in your engine, it can wear past the point of being reusable. The hex shaft that drives the pump should be given a close inspection. If you can feel wear at edges of the shiny portions of the shaft where it inserts in the bottom of the distributor and the top of the oil pump, replace it. Some engine rebuilders replace this component automatically because of its critical nature, however there's no need to replace yours indiscriminately. But on the other hand, don't take foolish chances. The shaft isn't that expensive. Remember, even though the oil-pump drive shaft may fail, the distributor continues to rotate permitting the engine to run, but the engine is not being lubricated. It doesn't take much imagination to visualize the resulting damage.

The oil-pump drive shaft for 351C and 351M/400 engines has a 5/16-inch hex 8-7/16-inches long—D0AZ-6A619-A. The 429/460 shaft has a 5/16-inch hex and is 9-11/32-inches long—C8SZ-6A618-A. If you decide to replace your drive shaft, don't discard it. Use it to check the length of the new one. It will also come in handy if you plan on rebuilding your distributor.

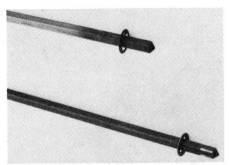

New oil-pump drive shaft above and old one below. Don't practice false economy by using your old drive shaft if its wear surfaces are more than just shiny. A failed shaft means instant oil-pressure loss.

Oil-Pump Damage—Before I get much further into the oil-pump subject, let's talk about durability. An oil pump is so over-lubricated that it's not going to wear much. Damage occurs when metal or dirt particles pass through the pump. This happens when some other component is chewed up—such as a cam lobe and lifter—or the oil filter clogs up and oil *by-passes* the filter, resulting in large dirt particles circulating through the oil pump and the engine. This can cause severe scoring and grooving of the oil-pump rotors and body—as well as bearing journals.

If your engine exhibits either condition, you may have already found oil-pump damage. Clues will have been given by the components you've already inspected and reconditioned, or replaced. Take this into account during your oil-pump inspection. For example, an engine which had heavily scuffed pistons and scored cylinder walls, wiped-out camshaft lobes and lifters or deeply grooved crankshaft bearing journals will have had large amounts of debris in the form of metal or dirt particles circulating through the oil pump.

Oil-Pump Inspection—Test your oil pump by immersing the pickup in a pan of clean solvent, then turn the rotors with your fingers using the hex shaft. After two or three complete turns, solvent should *gush* out of the pump's pressure port—the hole in the center of the pump-mounting flange. If the pump performs with solvent its internal clearances are not excessive, plus it shows the rotors turn freely.

Next, inspect the pump internally. Remove the four 1/4-20 bolts and the cover plate from the pump body. Be careful the rotors don't fall out. Use your feeler gages to determine the maximum clearance between the oil-pump housing and the outer rotor. It should not exceed 0.013 inch. Check the rotor end-play by laying the cover plate over half the housing and measuring clearance between the plate

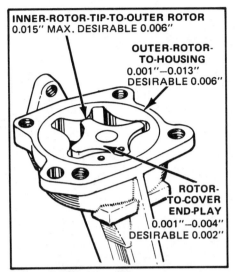

Use feeler gages to check oil-pump clearances to these specifications at the noted locations. I'm checking rotor-to-cover clearance. Other critical clearances are outer-rotor-to-housing and inner-rotor-tip-to-outer rotor. In addition to clearances, also check rotors and housing condition. They should be relatively free from scoring and grooving, particularly between housing's inlet and outlet ports.

and both the inner rotor and the outer rotor with your feeler gages. Maximum clearance is 0.004 inch. Next, remove the rotors. When turning the pump over, be ready to catch the rotors as they fall out. Be careful not to bang them around. Also, keep the rotors oiled to prevent rusting. Look at both rotors and the interior of the pump housing for signs of scoring, pitting or deep grooving, particularly between the inlet and outlet ports in the housing. A sure sign that the pump ingested foreign material is you'll find some imbedded in the tips of the inner rotor—another condition requiring replacement. If the rotors are damaged, they can be replaced for about a third of the cost of a new pump. Rotors have to be replaced in pairs and Ford offers replacement kit C8SZ-6608-A. Just make sure if you decide to do this that the surfaces in the pump body aren't damaged, otherwise little will be gained by installing new rotors.

Pressure-relief valve—The major concern with the oil-pressure relief valve is that it operates freely. To check for this, insert a small screwdriver into the *pressure port* and move the vale. If it moves freely against spring pressure, it's OK.

Assemble the Pump—Make sure the parts are clean. Install the rotors, then coat them liberally with oil. Make sure the outer rotor is installed correctly. If an *indent* mark is not showing similar to the one on the end of the inner rotor, turn the outer rotor over to expose its mark. Install the cover plate and torque the bolts 6–9 ft. lbs.

Before leaving the oil-pump subject,

If oil-pump pressure-relief valve moves free and easy when pushing on it like this, its OK. If it feels sticky and the rest of your pump is suitable for reuse, try freeing the valve and spring by soaking it in lacquer thinner or carburetor cleaner. If this frees it, the relief valve is OK.

I'll tell a little story to give you some ammunition for when you are trying to convince anyone of the dangers of driving a vehicle or operating any engine with little or no oil pressure. Any time an oil-pressure light comes on or a gage reads low pressure, an engine should be shut off *immediately*, not after you make it to the next gas station, another block, mile or whatever, but *right now*. A fellow I know had the engine in his pickup seize due to oil-pressure loss when the pump drive shaft failed—it broke. When I asked about whether or not he saw the *idiot light* he said, "Yes, but it wasn't bright red—just pink!" Using this reasoning, he tried to drive the last 10 miles home, but only made it five miles before the engine seized. The repair job came to approximately $500, or $100 per mile for the last five miles *after* he saw the light. This happened in 1962. What would the cost be with today's dollar?

BLOCK RECONDITIONING 73

6 | Head Reconditioning and Assembly

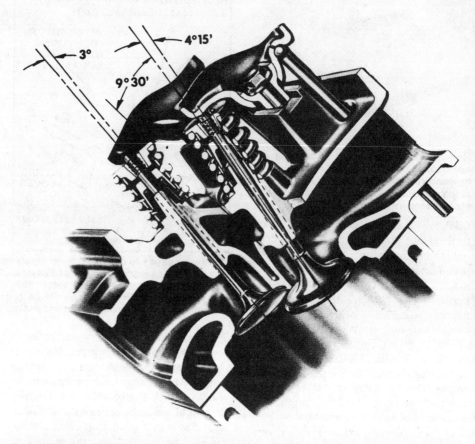

Pivot-guided, stamped-steel rocker arms used in all later 335 and 385 series engines. Drawing courtesy Ford.

I saved cylinder-head teardown and inspection until now because once you start disassembling your heads, you should continue working with them until the job is completed. This greatly reduces the possibility of losing or mixing pieces. Head reconditioning is a precise, tedious and dirty job requiring special tools and equipment. It's best to limit your head work to removing and installing unless you have the equipment *and* experience to do the job. You may do more harm than good if you attempt to do more. Leave the precision work to the specialist. It'll cost you no more for him to strip your heads too. However, if you have access to some or all of the equipment, by all means do it yourself. You can then inspect all the parts so you'll have an idea of what work needs doing to get your heads back into shape.

Rail-Rocker-to-Spring-Retainer Clearance— If you have a pre-1972 429, it will be equipped with the rail-type rocker arms. Check the clearance between the rocker-arm rails and the accompanying spring retainer. This dimension should not be less than 0.0625 inch (1/16 inch)—original clearance is about 0.080 inch. The reason this check is so important is it's possible the valve-stem tip and rocker arm will wear to a point where this clearance is reduced to nothing. Consequently, the rocker-arm rails begin to bear on the spring retainer, eventually causing the keepers to come loose. If this happens, the valve is free to drop into its cylinder, resulting in considerable engine damage. In fact, a dropped valve can literally destroy an engine, so be attentive when making this check.

Locate each rocker arm in its normal position on its *valve-stem tip* and measure the clearance between the rails and the spring retainer. A six-inch scale is handy for doing this, or use a 0.0625-inch-thick piece of sheet metal as a feeler gage. Discard any valve and rocker-arm combination under the 1/16-inch limit. The reason for replacing the rocker arm too is that excessive wear on the valve means the rocker arm is also too far gone—it will also have worn in an erratic pattern. So, if the old rocker arm were to be used with a new valve, excessive wear would probably result due to high valve-tip loading.

TEARDOWN AND CLEANUP

Keep the Rocker Arms and Their Fulcrums Together— When removing the rocker arms, their fulcrums and the attaching nuts or bolts, and possibly oil baffles, remember to *keep each fulcrum with its rocker arm*. String them on a piece of wire, less the ones you've found to be excessively worn. Just like camshaft lobes and lifters, rocker arms and fulcrums *wear in* together. Consequently, mixing them up will result in galling.

Adjustable and Non-Adjustable Rocker Arms— To repeat what I covered in the parts identification and interchange chapter, there are two basic types of rocker-arm setups—adjustable and non-adjustable. With one exception, the adjustable-type rocker arm is used only with a mechanical-camshaft-equipped engine—the 351C Boss/HO and 429 SCJ engines. The exception is the hydraulic-cam-equipped 429CJ which was originally equipped with adjustable rockers. This didn't last too long as they were changed to the non-adjustable type with positive-stop studs beginning with engines built after 10-31-69. Guide plates are used to stabilize the pushrods and rocker arms on all these engines.

Unlike the adjustable rocker arms, non-adjustable rocker arms are only used with hydraulic camshafts. The difference is in the type of rocker arm. The first type is the cast-iron rail-guided rocker arm used with a positive-stop stud as installed on '68–'71 429/460s. Next is the stamped-steel, barrel-type fulcrum secured to a notched rocker-arm stand by a bolt. This type rocker arm and fulcrum is found on all 351C-2V, -4V and CJ en-

If you have a pre-'72 429/460, check the clearance between the rocker-arm rails and their spring retainers. If valve tip and rocker arm are worn so rocker-arm rails are less than 1/16 inch from valve-spring retainer after valve's tip has been reconditioned, replace the valve. Also check rocker condition. If its valve-tip wear surface is badly worn, replace it too.

With a block of wood under the valve head and using a socket and a soft mallet, break spring retainers loose from keepers with a sharp rap on the socket. This makes compressing the springs an easier job and frequently dislodges the keeper to release the retainer, spring and valve.

Removing valves and spring assemblies from a head requires a spring compressor. Spring and its retainer are compressed just far enough so retainers can be removed from tip-end of valve, then released so retainer, spring, seal and valve can be removed.

What you're likely to find when inspecting your positive-stop rocker-arm nuts—cracks. Compare this new and old rocker-arm nut. Hair-line cracks on chamfer (arrow) of old nut means it must be discarded. Otherwise you run the risk of the nut cracking and releasing the rocker arm, its fulcrum and pushrod a short time after your engine is back in operation.

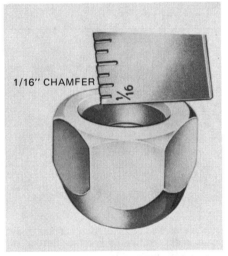
Inside chamfer of 429/460 positive-stop rocker-arm nuts should be 1/16-inch wide, otherwise nut will not position rocker arm at correct height. Check this dimension, even when using new nuts.

These valve-stem seals are in better condition than some I've seen, but are typical. Discard your old ones. New seals are included in a rebuild gasket set.

gines, all 351M/400s, '72 and later 429/460s, and '75 and later 460 Police engines. A cross between these two types of rocker setups is found on 429CJs built after 10-31-69, and 429 and '73–'74 460 Police engines. These use the same stamped-steel rocker arms as the 351C with a postive-stop stud and the Boss 351C-type fulcrum.

Positive-Stop Nuts—If you have a 429 or 460 using positive-stop studs, carefully check each nut for stress cracks. It's common for cracks to develop on the inside chamfer which bottoms against the stud shoulder. These cracks radiate outward from the threads, and are usually small and hard to see. So check them closely and replace those you find cracked. Don't hesitate to use a magnifying glass to check for cracks. If a nut with cracks is reinstalled, it is likely that it will split in two, releasing the rocker arm, its fulcrum and pushrod. One final point. Don't replace damaged nuts with just any old 3/8-24 nut, otherwise you'll find yourself in more trouble. Use Ford's C9SZ-6A529 nut or an equivalent OEM replacement nut.

Removing Valves and Springs—Now, you'll need the first piece of equipment not normally found in the standard tool chest—a *valve-spring compressor*. This is a specialized spring compressor for cylinder-head work. It compresses valve springs so the keepers, lock, collets, keys or whatever you want to call them can be removed or installed so the valve spring, spring retainer, valve and seal can be removed or installed.

There are two types of spring compressors, the *C-type* and the *fork type*. If you have a choice, pick the C-type. Fork-type compressors are typically used for removing and installing valve springs and

  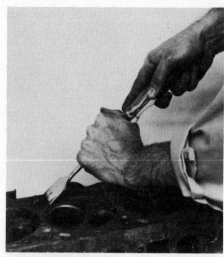

One valve can be saved and the other can't. Backside of intake valve is so badly carboned that air flow to its cylinder was reduced 20%. That means the engine's performance was also reduced. Because its stem wear was within acceptable limits, a good cleaning and refacing put it back into shape. Burned exhaust valve was discarded.

Get the gasket surfaces clean, right down to bare metal. This is necessary for good gasket sealing.

their components on heads still installed on their engines. The C-type compressor is easier to use on the *bench* because it straddles the head and pushes on the valve head and spring retainer simultaneously. This forces the spring into a compressed position while the valve stem projects out of the spring retainer so the keepers can be installed or removed.

Loosening Sticky Keepers—When compressing valve springs, you'll probably encounter sticky keepers. The trick here is to place the head right-side-up on your bench with a small block of wood under the head of the particular valve. Using a 9/16-inch, 3/8-inch-drive deep socket as a driver, place the socket squarely over the valve tip and on the retainer. A sharp rap on the socket with your soft-faced mallet will not only break loose the spring-retainer keepers, it may pop them out as well, releasing the retainer and spring. If you're lucky you won't have to use a spring compressor. If this method doesn't work 100%—it won't—you'll have to revert to the spring compressor and finish removing the keepers.

With the keepers out of the way and in a container where they won't get lost, you can release the pressure on the spring and remove the retainer and spring—if you're using a compressor. Slide the valve out of its seal and valve guide. This may not be as easy as it sounds. There's usually a burr around the tip of each valve which can cause a valve to hang up in its guide as it's being removed. This being the case, use a flat file to remove the burr by rotating the valve at its head while holding a file at an angle on the edge of the valve tip. To keep the valves in order, use

 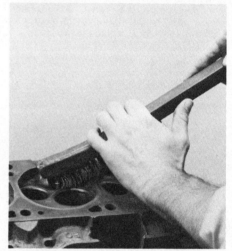

Remove all traces of carbon buildup from combustion chambers and intake and exhaust ports. A screwdriver is good for getting the big chunks and a wire brush gives the chambers the finishing touch.

Checking head-surface flatness with this precision bar and feeler gage. If a thicker than 0.006-inch-thick feeler gage fits between the bar and the head surface, resurface the head.

anything from a cardboard box to a piece of lath with 16 holes drilled in it.

CLEANUP AND INSPECTION

Scrape, Scrape, Scrape—Time for the old gasket scraper and elbow grease again. As you probably found when scraping your block deck surfaces, head gaskets seem to have "grown to" the metal. If you had your heads hot-tanked, the job will be much easier. Clean all gasket surfaces, the head-gasket, intake-manifold and rocker-arm-cover sealing surfaces of all old gasket material. Using a round file, clean the head-to-block water passages of all deposits.

It's time to *misuse* a screwdriver or a small chisel. Scrape the carbon deposits from the combustion chambers and exhaust and intake ports. After the heavy work is done, finish the job with a wire brush. This will put a light polish on the surface so you can spot any cracks which may have developed in the combustion chambers, particularly around the valve-seat area. If you do find cracks in any of these areas, don't junk the head just yet. Take it to your engine machinist to see if it can be repaired by *pinning*. If it can't, then it's junk. Even though it may be repairable by welding, chances are this cost will exceed the cost of replacement.

Head-Surface Flatness—Because cylinder heads are subjected to extreme heat, combustion and compression loads and are structurally weak compared to an engine block, they are susceptible to warping, as well as cracking. Excessive mismatch between block and deck surfaces causes combustion or coolant leakages, consequently those surfaces *must be flat*. Check the head-gasket surfaces with a *straight-edge* and feeler gages. Set the straight-edge lengthwise and diagonally across the head in both directions. Measure clearances which appear between the head and the straight-edge with your feeler gages.

The maximum allowable variation in the gasket surface is 0.003 inch over any six inches of length and 0.006 inch for the overall length of the head for engines with 10:1 compression or more; 0.007 inch for those with lower compression ratios. Limit warpage to half of the figures I just quoted if you plan on using *shim-type* head gaskets as opposed to the more compliant *composition* gaskets.

Cylinder-Head Milling—Even though one of your heads needs resurfacing, do them both. You don't want higher compression on one side of your engine to cause rough idling, and the cost is minimal for having the second head done.

Although more material can be removed from a cylinder head, I suggest you limit the amount removed to 0.010 inch. If necessary, you can remove up to 0.040 inch, however the problem with taking this much off is you'll boost your engine's compression ratio substantially—not a good idea with currently available fuels. This becomes particularly critical if your engine is one of the pre-1972 premium-fuel engines. Mill the heads on one of these engines and you will probably end up with an engine which can't be operated without detonation problems.

Another consideration when milling your cylinder heads is the intake manifold. It must also be milled if more than 0.020 inch is removed from the heads. Otherwise, the intake ports, water passages and bolt holes won't line up as they move down and toward the center of the engine. Therefore, the manifold bottom and cylinder-head gasket surfaces must also be milled. Less than 0.020 inch removed from the heads is not significant

Resurfacing a warped cylinder head. Accompanying chart shows intake-manifold milling requirements when more than 0.020 inch is removed from cylinder heads.

INTAKE-MANIFOLD VERSUS CYLINDER-HEAD MILLING REQUIREMENTS (INCH)		
Removed from Cylinder Heads	Removed from Manifold Bottom	*Remove from Manifold Sides
0.020	0.028	0.020
0.025	0.035	0.025
0.030	0.042	0.030
0.035	0.049	0.035
0.040	0.056	0.040
0.045	0.063	0.045
0.050	0.070	0.050
0.055	0.078	0.055
0.060	0.084	0.060

*Double amount when milling one side of manifold.

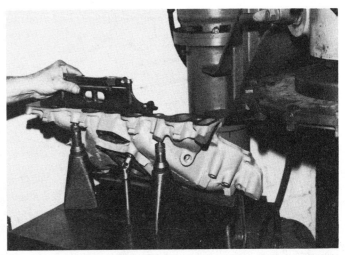

Setting up a 351C-2V intake manifold in preparation for milling right manifold-to-head mating surface. The bottom-side of the manifold has yet to be milled. Use chart to determine how much material should be removed from intake manifold relative to what's taken off the heads.

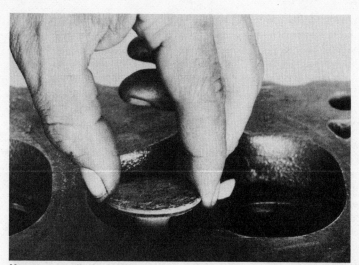 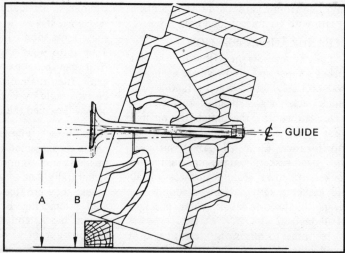

You can usually feel valve movement in a guide regardless of whether it is or isn't worn too much, consequently this is not a proper way of judging guide wear. A better check is to install a valve in the guide so its tip is even with the top of the guide. Measure how much it wiggles—the difference between A and B. Divide this figure by 3 to get the approximate stem-to-guide clearance. Measure the wiggle sideways too. Block the head up so the guide is parallel to the surface you are measuring from.

enough to cause head and manifold misalignment problems. A nearby chart provides machining requirements related to cylinder-head milling.

Rocker-Arm-Stud Replacement—Three defects require the replacement of a rocker-arm stud: damaged threads, a broken stud and one that is *notched*. Notching occurs when a rocker arm used with the ball-type fulcrum rocks over on its side, resulting in the edge of the clearance hole in the rocker arm contacting the stud. This wears a notch in the side of the stud, and could cause the stud to break. Replacing a stud is relatively simple. Unscrew the old one and thread in the new one. You'll need D0OZ-6A527-A if yours is the positive-stop type and C9ZZ-6A527-A for the adjustable type.

VALVE GUIDES AND VALVE STEMS

Valve-guide inspection and machining marks the beginning of hard-core cylinder-head machine-shop-type work which carries over into valve and valve-seat reconditioning. This work requires special equipment and special skills. If done incorrectly, it will ruin what would otherwide be a successful rebuild. If a valve guide is badly worn, it won't guide the valve so it can close *squarely* on its seat. Consequently, a valve can't seat properly as it will wiggle and bounce on the seat before closing. It may not even fully close at high RPM. This eventually *beats out* a valve seat and accelerates guide wear. Secondly, a worn guide lets too much oil pass between the valve stem and guide, resulting in excessive oil consumption.

With these points in mind, I'll cover the basics involved in reconditioning your seats, guides and valves.

Measuring Valve-Guide Wear—There are four ways to measure valve-guide wear: With a dial indicator at the top of a valve located in the guide, by *wiggling* the valve in the guide and measuring its movement, wih a *taper pilot* and a micrometer or with a *small-hole gage* and micrometer.

To check guide wear or clearance with a dial indicator, install a valve in the guide. Mount the dial indicator 90° to the valve stem in the direction you want to measure wear. Maximum wear usually occurs in the plane of the valve stem and the rocker-arm pivot. It can be in the opposite direction—in the plane running lengthwise with the head if rail rocker arms are used. With the valve about 1/8 inch off its seat and the indicator tip—it should be a flat one—contacting the valve stem close to the top of the guide, hold the stem away from the indicator and zero its dial. Push the tip-end of the valve stem toward the indicator and read the clearance directly.

Wiggling a valve in its guide to determine guide wear requires a minimum amount of equipment. It's not the most accurate method, but it's good enough to determine whether your guides need attention. It's done by measuring how much a valve wiggles, or moves at its head when pulled out of its guide. The amount of wiggle divided by 3 is *approximately* the stem-to-guide clearance of a *new valve* in the direction of movement. Measure valve wiggle with a dial indicator or the depth-gage end of a vernier caliper to be accurate. Using a taper pilot is the least dependable way of determining guide wear. A taper pilot is a tapered pin inserted into the end of a guide until it is snug. The diameter where the pilot stops is miked to determine guide diameter at its top or bottom. The measured diameter less the specified guide diameter is taken as the amount of guide wear—but it's not correct. The pilot measures the *minimum distance* across the guide rather than the *maximum distance*. It's the maximum distance you should be looking for, so don't use this method.

The most accurate method of determining guide wear is to use a small-hole-gage. You'll want the C-gage. Unfortunately, a small-hole-gage set costs about $30, and that's the only way you can buy them, in a set. However, if you have access to one, or have decided to make the investment, here's how to use it.

Insert the ball-end of the gage in the guide and expand it until it fits the guide with a *light drag*. Check the guide bore at several places up and down and around the bore to locate maximum wear. After setting the gage, withdraw it and mike the ball end. This will give you a *direct measurement* of the guide-bore cross section at the specific point you're checking. Subtract the valve-stem diameter from this figure and you'll have maximum stem-to-guide clearance. Measure around the guide at its top and bottom to find maximum wear. If the difference between maximum and minimum wear, or *out-of-round* exceeds 0.002 inch, your guides should be reconditioned.

How Much Guide Clearance?—Maximum stem-to-guide clearance is 0.005 inch. New stem-to-guide clearance range is

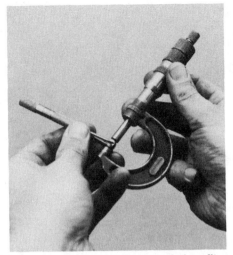

Measuring valve guides directly with a small-hole gage. Gage is first expanded to fit largest cross-section of guide, then gage is miked. Subtract 0.342 inch from the maximum gaged figure to arrive at guide wear. If it's 0.002 inch or more, the guides need reconditioning.

0.0010–0.0027 inch for the intakes and 0.0010–0.0030 inch for the exhausts. The limit you set for guide wear should not be the maximum 0.005-inch limit, but somewhere between the maximum figure and the *average* standard stem-to-guide clearance—an approximate average of 0.002 inch for both intake and exhaust. The number you decide on should be determined by the number of good miles you expect from your engine after the rebuild. By comparing the actual valve-guide clearance to the average standard clearance and the maximum wear limit, you can judge how many miles are left on your guides before the limit is exceeded. When doing this, keep one very important point in mind. As a valve stem and guide wear, they do so at an *increasing rate*. In other words, the more they are worn, the faster they wear. A guide and stem with 0.003-inch clearance may very well wear at twice the rate as one having 0.002-inch clearance. Consider the valve seats too. A valve that is sloppy in its guide will wobble when closing, resulting in the seat getting beat out. Incidently, valve-guide wear is higher in an engine used for stop-and-go driving, so all things considered, it takes good old "horse sense" when determining if it's time to redo your guides.

Valve-Guide Reconditioning—Valve guides can be restored by several different methods ranging from one I consider to be less than satisfactory to some that result in better-than-original valve guides. In the order of increasing desirability: knurling, oversized valves and guide inserts.

Guide Knurling—Knurling involves rolling a pattern into the existing guide to displace material and raise a pattern in the guide. This reduces the *effective diameter* of the guide, making it smaller than the original guide. By effective diameter, I mean the diameter inside the guide between the top of the displaced material, or *bumps* on one side, to the top of the *bumps* on the other side. Because the valve guide is now effectively smaller, it must be reamed to standard size. This smoothes off the peaks, or the top of the pattern created by knurling.

The problem with knurling is, it's the cheap way out. And as is usually the case, it's more cheap than effective. The reason is the actual contact area between the guide and the valve stem is substantially reduced, causing a proportionate increase in stem-to-guide contact pressure. Consequently, the guide and stem wear faster. Therefore, the oil-consumption problem caused by worn guides will not be cured, merely delayed. Don't use this method for reconditioning valve guides.

Oversize Valves—A fix which will restore stem-to-guide clearance is to ream the guides oversize and install valves with oversize stems. The obvious disadvantage is you'll have to purchase new valves, and they ain't cheap. Count on at least $210 for a complete set, plus a premium for the oversize models—and this does not include the cost of reaming the guides or the cost of the reamer if you intend to do it yourself. However, if some or most of your valves are not reusable, it may be a good way to go. You'll just have to make some price comparisons. Ford, and replacement-part manufacturers such as TRW, sell valves with oversize stems.

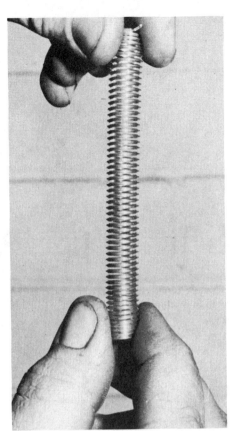

Screw-in bronze valve-guide insert. Guide must be tapped so guide can be threaded in. Once installed, insert is trimmed, locked in place and reamed to size.

Valve-Guide Inserts—Valve-guide inserts can restore the guides to as-good-as or better-than-original condition, depending on the type used. Two basic materials are used for guide inserts, cast iron and bronze. Cast-iron inserts restore the guides to original condition. They are installed by driving them into place after machining the original guide to a diameter equal to the OD of the insert, less a couple thousandths of an inch for an interference fit. This holds the new guide in place. After the guide is installed, it may or may not have to be reamed to size, depending on the way the insert is manufactured.

Similar to cast-iron guides, bronze and bronze-silicon guides are much more expensive as well as more durable. The bronze-silicon version is normally good for well over 150,000 miles. They are relatively easy on valve stems. These

CYLINDER-HEAD RECONDITIONING 79

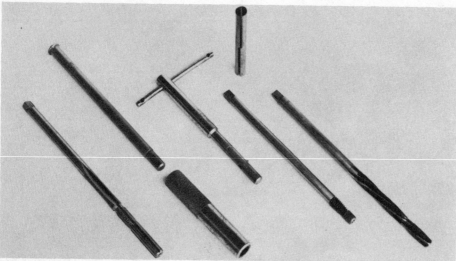

Thin-wall bronze valve-guide insert and tools required to install it. Guide is first reamed oversize so it'll accept bronze insert. Insert is then driven into place, trimmed, expanded, then finish-reamed.

Bronze-silicon valve-guide inserts are very durable. If you want to will your cylinder heads to your grandchildren, this is the type guide to install.

guides install in the same manner as cast-iron guides.

As for bronze inserts, they will provide better than original service. They come in two styles, the *thread-in* type and the *press-in* type. The thread-in insert is similar to a Heli-Coil replaceable thread except it's made from bronze with a thread form on its OD only as it's going to function as a valve guide rather than a thread.

To install the thread-type insert, a thread is tapped into the existing guide bore. A sharp tap must be used for doing this or the guide insert will not seat completely after being installed, and the insert will increasingly move up and down with the valve and actually pump oil into the combustion chamber. Consequently, this type of insert requires particular care when installing. After tapping, the insert is threaded into the guide and expanded to lock it into place so the valve and the insert can't work sideways in the thread, and so the backsides of the thread will be in total contact with the cast-iron cylinder head for maximum valve-stem-to-cylinder-head heat transfer, or cooling. The guide is reamed to complete the installation.

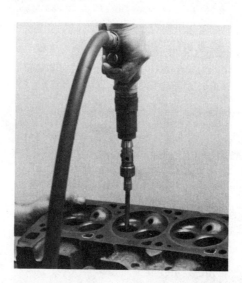

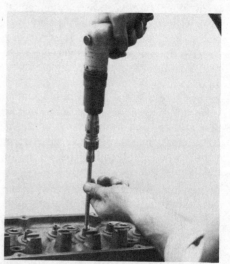

When installing a valve-guide insert, ream original guide oversize so it will accept insert. Next, this K-Line insert is driven into place with inner and outer mandrels. Excess insert material protruding from guide top is trimmed off, then insert is expanded so it fits tightly in original guide. Finally guide is reamed to size.

80   CYLINDER-HEAD RECONDITIONING

The K-Line bronze valve-guide insert is installed in a similar manner as the cast-iron insert. The existing guide is reamed oversize and the thin-wall insert—approximately 0.060-inch thick—is installed with a special driver. Once in place, the excess material is trimmed off. The guide is expanded and reamed to size using the *original* valve seat as a pilot to prevent *tilting* the valve guide. This should be done when reaming any type of guide.

VALVE INSPECTION AND RECONDITIONING

After getting your valve guides in shape, next on the list are the valves themselves. You should have already checked them for obvious damage such as burnt heads and excessively worn tips, particularly if your engine is the pre-'72 429 with rail rocker arms.

Measure Valve-Stem Wear—351C, 351M/400 and 429/460 engines use valves with nominal 11/32-inch stem diameters. More precisely, 351C and 351M/400 valve-stem diameters are 0.3416–0.3423 inch for the intakes and 0.3411–0.3418 inch for the exhausts. 429/460 intake and exhaust-valve stems are the same diameter at 0.3416–0.3423 inch. The reason for the smaller 351C and 351M/400 exhaust-valve stem is to provide the proper *hot* stem-to-guide clearance as all valve guides are reamed to the same diameter.

Use the preceding figures to check for stem wear, or you can compare the worn and unworn portion of each valve stem to determine exact wear. Maximum stem wear usually occurs at the tip end of a valve. This area is easily recognizable as the shiny portion of the valve stem. There's a sharp division between it and the unworn surface. Maximum valve opening is represented by the end of the shiny surface at the tip end. This is where a valve stops in its guide at its full-open position.

You'll need a 1-inch micrometer to check valve-stem wear. Measure the stem diameters immediately above and below the maximum-wear line, then subtract to determine wear.

Now that you've come up with a figure for valve-stem wear, the question is how much is acceptable? Again, this depends on the service you expect from your engine after the rebuild, if and how you reconditioned the valve guides and a myriad of other questions that makes arriving at an exact wear figure impossible. However, concentrating on desired service and the type of reconditioning you did on your guides, you should be able to answer this question.

Measuring valve-stem wear by comparing valve stem's worn and unworn portion. You'll find them directly below and above the wear line on valve's tip-end. This valve was replaced because I decided 0.0015 inch wear was too much for the additional mileage planned for the engine.

Reconditioning a valve. First grind the tip square to its stem—just enough to remove any signs of wear. Then chamfer tip to remove the sharp edge. Face is ground to a 44° angle to create a positive seal with 45° seat in head.

First, taking both ends of the spectrum, if valves with more than 0.002-inch wear are installed in guides with more than the 0.005-inch stem-clearance limit or in guides reconditioned by knurling, the time spent on head work to this point could've been saved by leaving the heads alone. Knurling is a Band-Aid fix rather than a true rebuilding method. It's like wrapping tape around a radiator hose to repair a leak instead of replacing the hose. The problem is not cured, it's just delayed. On the other hand, new valves or used ones with no more than 0.001-inch stem wear installed in guides reconditioned with bronze inserts should provide better-than-original service.

My suggestion is that you install valves with *no more* than 0.002-inch stem wear. Set this as your *absolute* minimum when deciding on which avenue to take when dealing with valve-stem and valve-guide wear. This should give you *at least half* the durability of new valves and guides.

Reconditioning Your Valves—Assuming your valves checked OK, have them reconditioned. Grind their *faces* and tips. The face is the valve surface which contacts the cylinder-head valve seat. A perfect seal must be made to seal the combustion chamber from the intake and exhaust ports when the valve is held closed

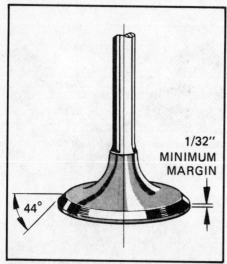

Exhaust valves should have no less than a 1/32-inch margin after grinding. Otherwise there is a risk of burning it. Intake-valve margins shouldn't be thinned beyond 1/64-inch.

A valve seat is ground to a 45° angle, then its outside and inside diameters are established by 30° top and 60° bottom cuts. A seat's outside diameter should be approximately 1/16-inch smaller than the outside diameter of the valve's face. Top and bottom cuts also establish seat width.

by the valve spring.

Valve faces are ground on a special grinder made just for this purpose. The valve is rotated in a *collet-type chuck* at an angle of 44° to a fast-turning grinding wheel. As the valve face rotates against the wheel's rotation, it is also oscillated across the face of the grinding wheel. Simultaneously, the valve and stone are bathed in cutting oil for cooling and to wash away the grindings. Enough face material is removed *only* to expose new metal on the valve face. If too much material is removed, the valve's *margin* will be thinned excessively and the valve will have to be junked. The margin of a valve is the thickness at its OD, at the outer edge of the face.

A valve with little or no margin will approach being sharp at its outer edge and must be replaced because it's highly susceptible to burning, particularly if it's an exhaust valve. An exhaust-valve margin should be at least 0.030 inch wide—approximately 1/32 inch, if you prefer fractions. Because an intake doesn't operate as hot as an exhaust valve, its margin can be as narrow as 0.015 inch—approximately 1/64 inch.

After the valve face is ground, the tip is faced, or ground square to the centerline of the valve stem, using another attachment on the valve-grinding machine. Finally, the tip is beveled, or *chamfered* to eliminate the sharp edge developed during the facing operation. Your valves should now be as good as new and ready for assembly into the cylinder heads—when the heads are ready. It's now unnecessary to keep your valves in order because all traces of mating to other components are gone.

VALVE-SEAT RECONDITIONING

Valve seats are reconditioned just like valves—by grinding. Equipment for valve seats is special, expensive and requires some degree of skill and experience to operate. This is a job for the specialist and it's an essential part of every rebuild. Seat reconditioning is particularly necessary for heads with reconditioned guides because the seats will not be *concentric* with the new guides—they won't have the same centers. Consequently, even though a valve seat may be in good condition, the guide won't let the valve seat evenly when it's closed. It will be held off-center in relation to the seat and cannot seat when fully closed. Because a valve seat is ground using a mandrel centered in the valve guide, the seat will be concentric with the guide.

Grinding the Seats—Valve seats are ground using a manual, electric- or air-powered grinder alternating between three grinding-stones with angles of 30°, 45° and 60°. Power tools are best because they provide the most accurate seat. The critical angle, 45°, is the valve-seat angle. If this 45° figure strikes you wrong, it's because the valves were ground to a 44° angle. The

Freshly dressed 60°, 45° and 30° stones ready to grind valve seats. I start with the 45° stone. After the seat cleans up—removing as little material as possible, zero in on seat width with the 30° and 60° stones.

one-degree difference provides an edge-contact at the outer periphery of the valve face so the combustion chamber will be sealed the first time the valve closes. Leakage due to poor seating can cause a warped valve, or a burned valve and seat.

The other two angles, 30° and 60°, are the *top cut* and the *bottom cut*, respectively. The 30° and 60° angles are important for low-valve-lift air flow. The 30° top cut is on the combustion-chamber side. It trues up and establishes the seat's outside diameter. The 60° bottom cut on the port side establishes seat ID, but more importantly, valve-seat width. Valve seat OD should be ground approximately 1/16-inch smaller than the valve OD. After the 30° top cut is made to establish seat OD, the valve seat is narrowed to the desired width with the 60° bottom cut.

Seat widths of 0.060–0.080 inch are used. For cooling reasons, exhaust-valve seats are generally ground to the high side of the range (wider) to provide additional valve-to-seat contact. This ensures additional contact area for valve-to-seat heat transfer.

Tools which should be on hand to assist in the grinding process, other than the actual grinding or machining equipment, are a set of dividers, Dykem-type metal bluing, a scale for setting the dividers for measuring seat OD and directly measuring valve-seat width and a special dial indicator for checking valve-seat concentricity. Bluing allows the seat to show up as a dark ring contrasted against the brightly finished top and bottom cuts, making the seat much easier to see and measure.

Hand-Lapping Valves—*Lapping valves in* is merely grinding a valve face and its seat *together* with *lapping compound*. This compound or paste is applied to the valve face, then the valve is rotated back-and-forth in a circular motion while simultaneously applying light pressure to the head of the valve.

I'm mentioning valve lapping in case you are aware of the process and think it is necessary for valve sealing. Generally speaking, if a valve face and its seat aren't ground correctly, lapping won't fix the problem, and if they are done right, lapping shouldn't be necessary for proper valve seating and sealing. So don't waste your time with lapping. A reputable rebuilder is your best guarantee.

If you do lap in the valves, be sure to clean off *all* the lapping compound. Grinding compound and engines do not make good traveling companions. When you finish lapping each valve, mark it so you can put it back on the same seat.

Do They Seal?—It would be nice to know if your valves are going to seal before installing the heads on your engine. This is so you can do something about it now if there is a problem. To determine if your valves are sealing OK, you'll first have to assemble the valves in the heads. Either do it temporarily now, then disassemble the heads for checking and setting up the valve springs, or complete the valve-spring checking and setting-up as outlined in the next few pages, then assemble the valves in the heads and check them.

With the valves in place, position the heads upside down on your bench so the head-gasket surface is level. Using kerosene, fill each combustion chamber and check inside the ports for leaks. If you don't see any you can consider the valves

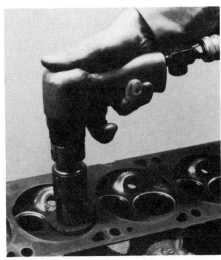

After valve-guide pilot is in place, the valve-grinding mandrel with its stone is slipped over it. Mandrel and stone are rotated with an air or electric tool.

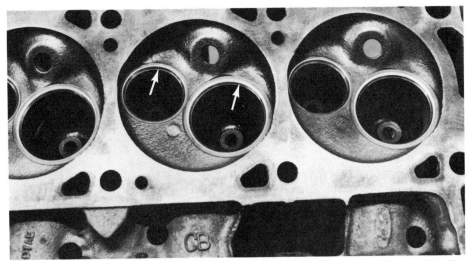

Dark bands (arrows) are valve seats. After finishing 45° cuts, seats are blued so they'll stand out from the 30° and 60° cuts for ease of measurement. A pair of dividers and a 6-inch scale are all you'll need for measuring.

After a valve seat is ground, its concentricity or runout in relation to the valve guide should be checked with a gage such as this. Photo courtesy Ford.

CYLINDER-HEAD RECONDITIONING 83

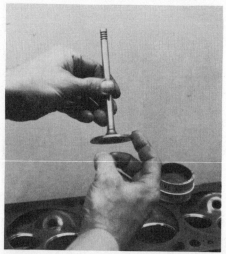

Lapping valves is unnecessary because it won't correct an incorrectly ground valve or seat, and it's not necessary if they are done right. It will, however show up a bad valve job. Lapping paste is applied to valve face. Valve is inserted in guide, lapping tool attached to valve head, then valve is rotated back-and-forth while being held lightly against its seat.

Lapped valve is one with grey band on its face which coincides with its seat contact area. If this band is continuous around the valve face and seat, consider your valve-grinding job OK.

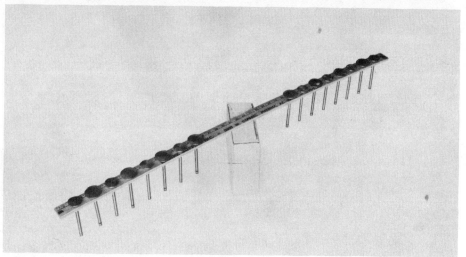

Valves should be installed on the seat they were lapped on. Keep them in order like I'm doing here, but don't lose track of which head they go with.

OK, otherwise you'll have to confer with your rebuilder to determine what the sealing problem is, then have it corrected.

VALVE SPRINGS—INSPECTION & INSTALLATION

After your heads and valves have been completely reconditioned, the valve springs must be inspected. A valve-spring tester helps, but it's not absolutely necessary. It's helpful to understand what's involved in the valve-spring testing. You should become familiar with the valve-spring terms and the importance of maintaining certain standards. The common terms are: *spring rate, free height, load at installed height, load at open height* and *solid height*.

Spring Rate—Spring rate is not one of the commonly listed valve-spring specifications as such, but it relates directly to most of the other specifications. Also, it's one of the basic terms necessary to describe a coil spring's mechanical properties.

Spring rate governs the load exerted by a spring when it is compressed a given amount. Rate is usually expressed in so many pounds-per-inch of deflection. When a spring is compressed some amount, it requires a given force to do so. Typical valve-spring rates vary between 300 and 450 pounds per inch. Spring rates are usually indicative of the RPM range of the engine in which they are installed. For example, the lightest spring in the 351C-2V has a 325 lbs./in. rate whereas the higher revving, solid-liftered 429SCJ has a valve-spring rate of nearly 450 lbs./in. A higher spring rate, which translates into higher installed and open loads, is necessary to control the additional valve-train inertia generated at high engine RPM. Increased spring loads are required just to close the valves, or keep them from *floating* when an engine is operated at high RPM. If a valve spring loses its rate, or *resiliency*, it is no longer capable of closing its valve at or near maximum rated RPM.

Free Height—Free height or length of a valve spring is its unloaded height or length. If a spring's free height is too long or too short, the load exerted by the spring as installed in an engine will be incorrect because it will be compressed more or less by the spring retainer. Therefore, knowing what spring rate is in addition to free height gives you a clue as to why a spring does or doesn't meet its specifications.

Load at Installed Height—A spring's load at its installed height is a common spring specification—installed height is spring height as installed in the cylinder head: measured with the valve closed. This is the distance from the cylinder-head spring seat to the underside of the spring retainer. A typical load-at-installed-height specification is 76–84 lbs. at 1.810 inches. When compressed to a height of 1.810 inches from its free height, the load required should be between 76 and 84 pounds. The absolute minimum installed load is 10% less than the lower load limit, or 68 lbs. in this case. If a spring does not exceed or at least meet the minimum, it should be replaced. If your engine will be operated at its upper RPM limit, or is one of the high-performance models, you should make sure valve-spring load at least meets the *minimum* standard limit.

Load at Open Height—Another commonly listed specification is a spring's compressed load when its valve is fully open. A typical specification is 240–266 lbs. at 1.330 inches. Just as with the installed-height load, a spring must fall within the 10% minimum-load limit or be replaced. In this case, the minimum is 216 lbs. Again, consider how your engine will be operated when checking the springs.

Solid Height—Solid height describes the height of a coil spring when it is totally compressed to the point where each coil touches the adjacent coil. The spring is said to *coil bind*, or *go solid*. Coil springs should never be compresed to this height in normal service. If a valve spring were to reach its solid height before its valve is fully opened, the load on the valve train would theoretically approach infinity. But before this could happen, the weakest component in the valve train will fail—something bends or breaks—usually a pushrod.

Squareness—Valve-spring *squareness* is how straight a spring stands on a flat surface, or how much it *tilts*. It is desirable for a spring to be square so it loads the spring retainer evenly around its full circumference. Uneven retainer loading increases stem and guide wear—something you should minimize. Limit out-of-squareness to 1/16 inch measured from the top of the spring to a vertical surface with the spring sitting on a horizontal surface.

CHECK YOUR VALVE SPRINGS

Of the spring characteristics just covered, the ones which should be checked are squareness, and the installed and open spring loads, plus making sure the spring doesn't reach its solid position in the full-open position. A flat surface, square and something to measure with are all that's required to check for squareness. When you're all set up, rotate the spring you're checking against the square to determine its maximum tilt, then measure it. If the gap at the top of the spring exceeds 1/16 inch, replace it.

As for the last two checks, a valve-spring tester is required. Two types of testers are normally used, but both do the same thing. They permit manual compression of a spring to the opened and installed heights so the spring's loads at these heights can be read. The question is,

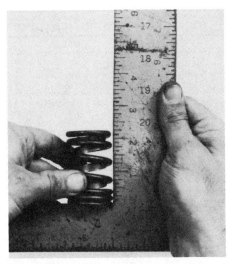 

Two quick and easy ways of determining whether your valve springs need shimming and/or replacing. Squareness is being checked with a carpenter's framing square. Spring should lean no more than 1/16 inch at the top. If you find free heights of your springs varying as much as these two, you'll know more comprehensive checking is needed.

Checking spring load at installed height. Shimming brings spring load within specification, then spring is checked at open height to ensure it doesn't go solid. You should be able to see daylight between each coil at open height.

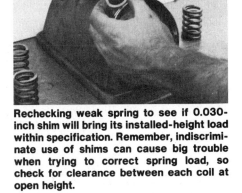

Rechecking weak spring to see if 0.030-inch shim will bring its installed-height load within specification. Remember, indiscriminate use of shims can cause big trouble when trying to correct spring load, so check for clearance between each coil at open height.

"How do I check for spring load if I don't have a spring tester?" The answer is simple, you don't!

Fortunately, a simple method can be used to determine if the spring is usable: compare its free height to that originally specified. A spring which has been fatigued or overheated, usually from excessive engine temperatures, will *sag* or collapse some amount. This is reflected in a spring's free height. A sagged spring's load is reduced in the installed and open positions, or heights.

Also, when a spring is removed from

CYLINDER-HEAD RECONDITIONING 85

VALVE SPRING SPECIFICATIONS

Engine	Year	INSTALLED Height (in.)	INSTALLED Load (lbs.)	INSTALLED Min. Load (lbs.)	OPEN Height (in.)	OPEN Load (lbs.)	OPEN Min. Load (lbs.)	Free Height (in.)	Solid Height (in.)	Color Code
351-2V	70-71	1.820	76-84	68	1.420	199-221	179	2.07	1.37	Yellow or 3 Yellow Stripes
	71-74	1.820	76-84	68	1.390	215-237	183	2.06	1.33	White or 2 White Stripes
351C-4V CJ	70-74	1.820	85-95	77	1.320	271-299	244	2.05	1.26	2 Lt. Green Stripes & 1 White Stripe
351 Boss 351C H.O.	71 72	1.820	88-96	79	1.320	299-331	269	2.03	1.27	4 Lt. Blue Stripes or 3 Green Stripes
351M/400	75-77	1.820	76-84	68	1.390	215-237	183	2.06	1.33	All White or 2 White Stripes/1 Red & 1 Silver Stripe
	78-79	1.820/1.68	76-84/80-88	68/72	1390/1.25	215-237	183	2.06/1.93	1.33/1.20	
429/460	70	1.810	76-84	68	1.330	240-266	216	2.03	1.28	Blue
	71-72	1.810	76-84	68	1.330	217-241	195	2.07	1.27	3 Gold Stripes
	73	1.780	71-79	66	1.390	161-177	145	2.09	1.30	Black
	74	1.810	76-84	68	1.330	240-266	216	2.03	1.28	Blue or 2 Blue Stripes
	75-79	1.810	76-84	68	1.330	217-241	195	2.07	1.27	2 or 3 Gold Stripes
429CJ, SCJ Police	70-72	1.820	88-96	80	1.320	300-330	270	2.03	1.27	4 Lt. Blue Stripes
460 Police	73-74	1.820	88-96	80	1.320	300-330	270	2.03	1.27	4 Lt. Blue Stripes
	75-76	1.810	76-84	68	1.330	240-265	216	2.03	1.28	Blue or 2 Blue Stripes
	77-78	1.820	88-96	68	1.320	300-330	270	2.03	1.27	4 Lt. Blue Stripes

If specifications for intake- and exhaust-valve springs are different, intake spring will be shown first (intake/exhaust).

its cylinder head, it will not return to its original, or specified free height. So, use the valve-spring chart to determine what the free height of your springs should be, then measure them. Generally, a valve spring should be within 0.0625 inch of the specified height if it is to be reused in your engine.

Valve Spring—Replace or Not to Replace?— I replace any spring which is 1/8 inch shorter than its specified free height. A valve spring that's 1/16 inch shorter can be installed in an engine that's to be used for light service. This decision should be based on what type of service your engine will see and how far off your springs are. So, if it's going to be used for puttering around town, you probably can get by with marginal springs. This is not the case though if the engine is to be operated at or near its peak RPM. This being the case, the springs will get progressively weaker and those valves with the weak springs will float, limiting engine RPM to a lower speed than you might want.

If your engine is to be used for severe service, the springs should be tested on a spring tester to determine if they can be corrected by *shimming*, or placing a shim or *spacer* under the weak springs, or those that have a *short* free height. In this application, the shim—a special flat washer—is placed between the spring and the cylinder head so the spring will be compressed more to restore the spring's installed and open loads. However, if not used carefully, shims can cause severe valve-train damage by causing the spring to reach its solid height before its valve is fully opened. Also, another problem when using valve-spring shims is a spring is designed to be compressed a certain amount *continuously*. This amount is the difference between its free height and open height. If it's compressed more than this amount, the spring will be over-stressed, resulting in an over-worked, or fatigued spring. The spring will quickly lose its load-producing capability. So if you have a questionable spring, replace it rather

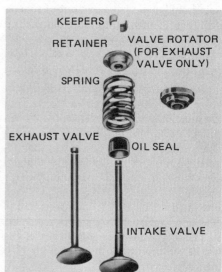

Components for valve-and-spring installation. One-piece spring retainer is shown here, however two-piece retainers are used in some instances as are valve rotators on exhaust valves. Your spring assembly may also include a damper, a flat-wound spring which fits inside the spring. Photo courtesy Ford.

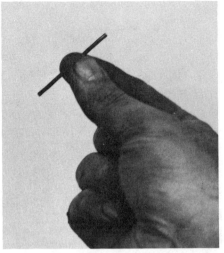

Using a short section of wire cut and filed to match spring-installed height to check seat-to-retainer distance. If this distance is too much, make up the difference up with a shim and recheck using the wire "gage." When you've found the right combination, keep the valve and shim together so you can install them in the same position as fitted.

If your gasket set includes the hard nylon valve-stem seals rather than the less-durable soft polyacrylic seals, you'll have to use the installation tool included in your gasket set. Each seal is literally driven onto its stem while the valve is supported under its head.

than shimming it, even though a new spring costs more than shims. If you do shim some of your springs, make sure you attach the shim to its spring with some wire, string or whatever. If they get mixed up, you'll have to retrace your steps and recheck them.

ASSEMBLE YOUR CYLINDER HEADS

Now that all your cylinder-head parts have been inspected and/or reconditioned, you can get them all back together again—sort of like Humpty Dumpty. Except in this case you won't need the King's Horses or Men, just a valve-spring compressor and a ruler, plus some valve-spring shims—maybe. Yes, you may still have to install shims even though you may have gotten by without them up to this point.
Valve-Spring Installed Height—One of the valve-spring loads you checked was the spring's *installed height*, or the height the valve spring should be with the valve closed and the spring and its valve, retainer and shim/s installed. Shims are included in the spring's installed height *if used to correct spring free height*. Make sure every spring is at its correct installed height, otherwise its valve will not be loaded correctly.

A valve stem projects out the top of the guide when the valve is closed. The distance from the cylinder-head spring seat to the underside of the valve-spring retainer installed on the valve, is also determined by the valve seat.

As originally manufactured, this distance, or pad-to-retainer dimension, coincides with the valve-spring installed height. However, because material is removed from each valve face and seat during the grinding process, this distance is increased and consequently exceeds the installed spring height. If your valve seats were in good condition, only a small amount of material should have been removed to clean them up. Therefore, the spring's installed height should not have been affected to any great degree. On the other hand, if more material was removed, or the heads have been rebuilt before, the spring's installed height and load could be changed considerably, so they should be checked.

Because setting and checking installed spring height is best done as part of the cylinder-head assembly process, let's get on with putting your heads together and do the height checks at the same time. Have your valves ready to be installed in your heads *in order* and the springs with their shims, if you used any to correct valve-spring free height.

Installed spring height can be checked by two methods. You can simply insert the valve in the head, then assemble the retainer and keepers on the valve stem without the spring, shims or valve-stem seal. The retainer is lifted up to hold the valve in its closed position—it also keeps the retainer and keepers from falling down the stem—while the seat-to-retainer dimension is measured. To get an accurate reading, the retainer must be lifted up firmly to ensure the valve is against its seat and the retainer and keeper are in place. A *snap-gage* and micrometers will give you the most accurate reading when making this measurement, however you can do pretty well with a six-inch scale if you have a sharp eye. A neat trick to use is to cut, then file or grind a short section of welding rod or heavy wire to the specified installed height of the spring. You can accurately check the rod length with a micrometer. This will give you an accurate gage which when placed between the spring seat and the spring retainer will show a gap the same thickness of the shim required. The thickest shim that best fits between the rod, retainer and spring seat is the one to use.

If you've used the measuring method to determine shim thickness, after you've arrived at a shim-pad-to-retainer figure, subtract the installed spring height from it and you have the shim thickness required to provide the correct installed height. Therefore:

Shim thickness = spring-to-retainer distance - installed spring height

Shims come in thicknesses of 0.015, 0.030 and 0.060 inch. They can be stacked in varying combinations so you'll end up with the right overall thickness.

The second way to check seat-to-retainer distance varies only in that you install the spring too. This means you'll have to use the spring compressor to do so, but it does ensure the reading will be accurate, just more difficult to take. You won't be able to use a *snap-gage* or anything between the retainer and the spring seat. One measuring device that works well here is a pair of vernier calipers. You can also use your less-accurate six-inch scale. Measure the actual spring height in this case by measuring from the spring pad to the top of the spring. Regardless of the method used, the way of arriving at shim thickness is the same—subtract *specified*

After oiling the valve stem and its guide, slide valve into the guide and install valve-stem seal. This soft-type seal is easy to install. Push seal down over valve stem so it'll hold valve closed. Your hands are then free to install the spring, retainer, keepers and any shims. Note valves lined up in order with shims, springs and retainers slipped over valve stems. This will minimize installation errors.

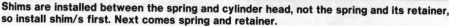

Shims are installed between the spring and cylinder head, not the spring and its retainer, so install shim/s first. Next comes spring and retainer.

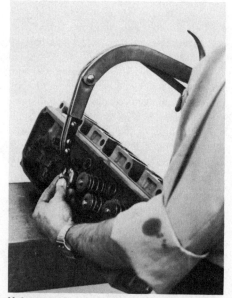

Using a spring compressor, compress valve spring only enough to install keepers. Grease on keepers will hold them in place while you release compressor. Check keepers to make sure they are firmly in place.

installed spring height from the dimension you just measured. If the variation is less than 0.015 inch, you don't need a shim.

When checking installed spring heights, keep a record of the shims and their location as you go along. Or better yet, line up the valves with their respective shims, springs and retainers over the valve stem in the exact order they will be installed in the head. You can just go down the line when doing your assembly. Remember, you can mix up seals, retainers and keepers, but the *springs, shims and valves must be installed in the position in which they were checked.*

After finishing the installed-height check, use your list to purchase the necessary shims if you haven't already done so. Also, before assembling your heads, make sure all the parts are clean. Using your shim list, organize the shims you just purchased so they'll be in the right location for installation time. One thing that's handy to know when looking for a particular thickness shim is they are sometimes color-coded according to their thickness. For example, V.S.I. uses the following color codes: Black, 0.015 inch; silver, 0.030 inch; gold, 0.060 inch.

Install the Valves—Set your heads on their sides on your work bench and start assembling at one end of a head with the valves, springs and shims lined up in order. Confusion is minimized by working in the same order you used when checking installed spring heights. Oil each guide and valve stem, insert the valve in its guide, and slide a stem seal all the way down the valve stem. This will keep the valve from falling out of its guide so you can concentrate on installing the shim/s, spring, damper spring, if your engine is so equipped, and spring retainer in that order. Stem seals are included in your engine's gasket set.

Two types of valve-stem seals are used, either of which may be included in your gasket set depending on which engine you have. The first, and most common, is the soft polyacrylic type which is easy to install by simply working it over the valve-stem tip, then pushing it down over the stem by hand. The other, and more difficult to install seal, is a rigid or hard nylon seal. If your gasket set includes this kind of seal, it will also include an *installation tool*. This tool is a plastic tube which bears against the seal for driving it down over the valve stem with a hammer. When doing this, the valve must be supported squarely under its head to eliminate the chance of bending the valve stem. Wrap Scotch tape over the keeper grooves to avoid seal damage. Lubricate the valve stem so the seal will install easier.

After the valve and seal is in place, the shims, spring—and spring damper in some cases—spring retainer and the keepers are next in that order. It's possible that you'll have shims for correcting spring *installed load* as well as for *installed height*—or you may have none at all.

At this point, use the spring compressor to pull the retainer down around the valve-stem tip for installing the keepers. Don't compress the springs more than necessary. With the keepers in place, release the spring compressor and check the keepers to make sure they stay in place. Correct them if they are not. Continue installing the valves and springs. If you made a wire gage, use it now to confirm

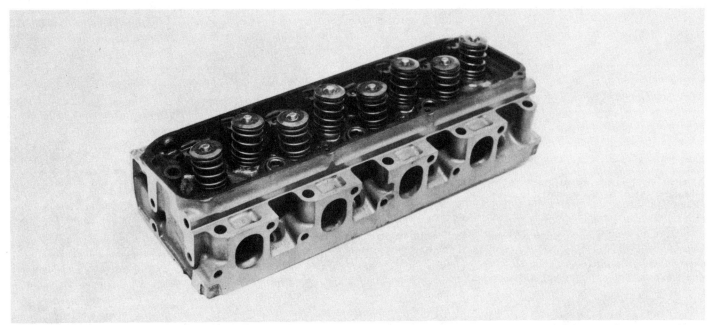

Except for rocker arms and fulcrums, this is a complete cylinder-head assembly. I prefer installing rocker arms after heads are on engine. So, for now I cover the heads up until engine is ready.

installed spring height.

With the installation of the last spring and valve, your heads are assembled to the point that they can be installed on your engine block when it's ready. Spray the heads with oil and cover them so they'll remain clean, dry and rust-free until you're ready for them.

INTAKE MANIFOLD

All that's required to restore your intake manifold to like-new condition is a thorough cleaning. If it was hot-tanked, this job will be considerably easier. Scrape its gasket surfaces—head, cylinder block and carburetor.

Remove the Baffle—Early 429/460 engine intake manifolds have a sheet-metal baffle riveted to their undersides to shield engine oil from the hot underside of the heat passage. This type of baffle was used rather than the later type which is an integral part of the intake-manifold-to-cylinder-head gaskets. To clean the sludge build-up from this area, you'll have to remove the baffle. Two spiral-groove rivets secure the baffle to the manifold. To remove them, you'll need to wedge something with a very sharp wedge under the rivet heads. A long skinny chisel will do. One thing to be aware of, the rivet-heads are easily broken off. If this happens, drill out the remainder of the rivet.

Wedge under the rivet-head until the head is about a quarter inch off the baffle— far enough to clamp onto the head with a pair of Vise-Grip pliers. With the pliers clamped on the rivet head, pry under the pliers while turning the rivet counter-clockwise to unscrew it. When you've

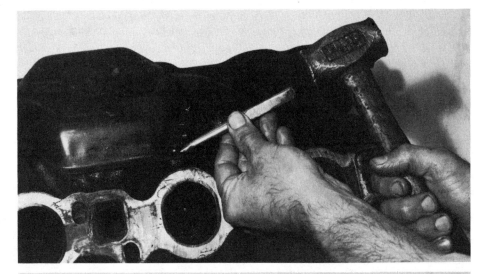

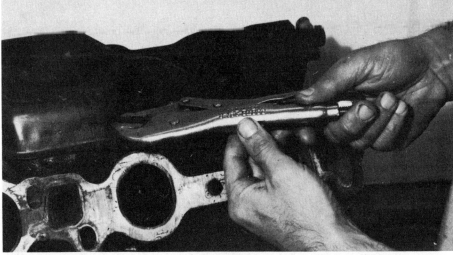

To do a complete job of cleaning an early 429/460 intake manifold, you'll have to remove the heat-riser baffle. Do this by wedging under the rivets to raise them far enough so you can clamp on with a pair of Vise-Grip pliers so they can be "unscrewed" from the manifold. Be careful, their heads are easily broken off.

A baffle rivet has a spiral "thread" for retention. Pitch is coarse so it can be driven into place.

removed the first rivet, back out the second one just far enough so you can rotate the baffle out of the way so you can remove the sludge and caked-on deposits.

After you have the manifold clean, rotate the baffle back into place and install the rivet. Here's where skinny fingers or needle-nose pliers come in handy. Hold the rivet so it starts in squarely, then tap it lightly into place.

Break a Rivet?—It's relatively easy to break off a rivet head, but don't be discouraged if you do. Get on with the job of removing the remainder of the rivet so you can reinstall the baffle. Don't think you can do without the baffle because if you reinstall the manifold without it you'll end up with an oil-consumption problem due to the oil atomizing when it comes in contact with the hot underside of the intake manifold. It will then go out the engine's PCV system.

Rather than trying to find a replacement rivet, replace it with a 1/4-20x3/8-inch bolt. First, drill an 1/8-inch pilot hole in the center of the broken rivet. Center-punch the rivet first to make sure the drill doesn't run off center. Next, use a 7/32-inch drill to remove the remainder of the rivet. A 7/32-inch diameter hole is the correct size for a 1/4-20 tap. After tapping the hole and cleaning the manifold, reinstall the baffle using the bolt and a lockwasher.

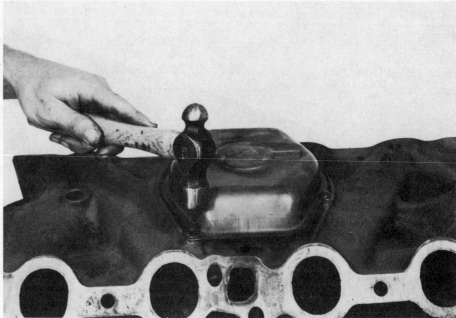

After sludge is cleaned from underside of intake manifold and inside baffle, reinstall baffle and its rivets. Lightly tap rivets into place.

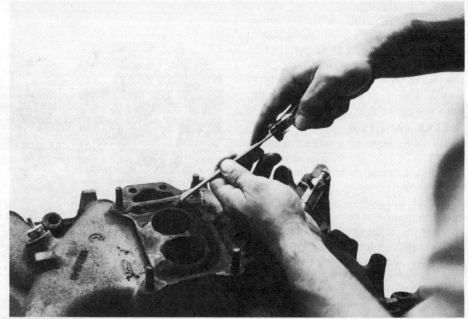

Don't forget heat-riser passages. Remove carbon deposits so your newly rebuilt engine will have quick warmup after start-up.

7 | Engine Assembly

This is the part of engine building I like. Everything is clean, all the parts are new, reconditioned or have been checked out. The running around associated with getting parts and jobs done you couldn't handle is just about over.

Things You Need Before Starting—Just like all the jobs you've done until now, there are things you'll need for assembling your engine in addition to all the parts. Trash bags are useful for covering up clean parts, particularly the block, during assembly. You'll aso need a complete gasket set, gasket sealer, gasket or weatherstrip adhesive and a spray can of aluminum paint. All sorts of sealers and adhesives are available, but I'll list a few that work particularly well in certain applications. First is room-temperature-vulcanizing (RTV) silicone sealer. It's great if it's used right in the right place and in the right amount. Used incorrectly, it can be disastrous. You have to know its limitations. It's not a cure-all. Ford markets some fine sealers. Perfect Sealing Compound, B5A-19554-A, is a general-purpose sealer. Gasket and Seal Contact Adhesive, D7AZ-19B508-A is especially good for installing intake-manifold gaskets. Another good one for this purpose is OMC's (Outboard Marine Corporation) Adhesive Type M. You can also use weatherstrip adhesive as a gasket-adhesive: Ford's C0AZ-19552-A or 3M Corporation's 08001.

Lubricants are a necessity when assembling an engine. How an engine's critical parts are lubricated during its first few minutes of initial run-in will be a major determining factor in the engine's durability. Remember this during the assembly process. Lubricants to have on hand include at least a quart of the oil you intend to use in your engine's crankcase—probably a multi-grade detergent type—a can of oil additive and some molybdenum-disulfide grease. As for what brand of oil to use, I won't make any suggestions because the brand isn't as important as the grade. So, regardless of the brand you choose, use the *SE grade*. In addition to crankcase oil, get a couple of cans of Ford's Oil Conditioner, D2AZ-19579-A, or GM's EOS (Engine Oil Supplement) for general engine assembly,

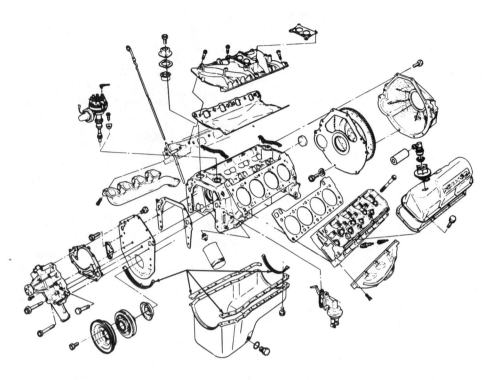

351C external components. Although there are variations, particularly front-cover designs, 335 and 385 Series engines appear basically the same. Drawing courtesy Ford.

initial bearing lubrication, and to put in the first crankcase fill. Finally, in the oil department, you could also use a squirt can. Fill it with motor oil for ease of application.

Tools—Now, for special tools other than those normally residing in your toolbox. You must have a torque wrench because *an engine cannot be assembled correctly without a torque wrench*. When it comes to tightening bolts, even the most experienced mechanic doesn't rely on *feel*, he uses a torque wrench. Therefore, put this tool at the top of your list.

The next item is one that I don't consider necessary because bearings are made to such close tolerances, however it's not a bad idea to use *Plastigage* as a check to be sure you get the right bearings. It's not uncommon for the *wrong bearings* to be in the *right box*. Plastigage is a colored strip of wax used for checking assembled bearing-to-journal clearances. All you need to know about it at this point is you'll need the *green* Plastigage which measures 0.001–0.003-inch clearances. Get one envelope or strip of it—you won't need much.

The need for the next tool depends on the route you took with your cam bearings. If you left your old ones in the block or had your engine machinist replace them for you, you won't have to concern yourself. If you have yet to replace them, you will need a cam-bearing installation tool set. Before you rush out to rustle one up I'll suggest you have an engine shop do it for you. It's not too late yet. However, if you still insist on doing it, read on.

CAM BEARING INSTALLATION

Guess I didn't scare you out of attempting to replace your own cam bearings. Well, you can't back out now, so get on with it. The first thing you'll need are the tools. There are two kinds: one pulls the bearing into its bore with a threaded rod, nut, thrust bearing and a *mandrel*; the other drives the bearings into place with a mandrel, drive bar and a hammer. When installing cam bearings, three things must be kept in mind. First, the bearings must be installed square in their bores, oil holes in the bearings must line up with those in the block and care must be taken so as not to damage the bearings.

Different Sizes—The 351Cs and 351M/400s have progressively smaller camshaft-bearing inserts from front-to-back: 2.124,

2.066, 2.051, 2.036, 2.021 inches, respectively. Because the bearing shells are the same thickness, the 351C and 351M/400 bearings must be installed in the right bearings bores. If you attempt to install one in the wrong bore, it will fit loosely, or be destroyed. The 429/460 bearing journals all have a 2.124-inch diameter.

Chamfering the Bearings—Before installing your cam bearings, it's a good idea to remove the sharp edges from the inside edges of new bearings. They can interfere with the cam journals as you're trying to slide the cam into place, making cam installation more difficult than it needs to be. The tool particularly suited for this job is a *bearing scraper*. It's like a triangular file with no teeth. If you don't have one of these, use a pen knife.

To chamfer the bearing ID, hold your knife or scraper 45° to the bearing surface and hold the bearing so you can rotate it while peeling a small shaving about the size of four human hairs—if you can picture the size of four hairs. The idea here is to remove just enough so you can't feel a burr when you drag your fingernail across the edges of the bearings.

Get the Block and Bearings Ready—Put your cylinder block on its back so you'll have access for installing the bearing inserts. Also, it's best to locate the block so you can easily sight down the center of the bearing bores during the installation, particularly if you have a bearing installer that's not self-centering.

After your block is positioned, organize the bearings in the same sequence they are to be installed in. This will save some time fumbling around. Also, fully understand where the bearing-insert oil holes are to be positioned in their bores so they will coincide with the oil holes in the block. Cam-bearing lubrication depends on oil fed from the crankshaft main bearings. If an oil hole is closed off due to a wrongly positioned bearing insert, goodbye bearing!

Bearing-insert oil holes are slotted to accommodate some misalignment with the block oil holes, so they don't have to line up exactly, but this is no excuse to get careless. Another thing to watch is that the number-one bearing bore has two oil holes. One feeds the cam journal from the crankshaft and the other feeds the distributor journal from the cam. So there are two holes to line up in the number-one position. Because all the 429/460 bearing inserts are the same, they have an extra oil hole which is not used in the other four bores.

Two excellent engine-assembly items. Comparable aftermarket sealers and oil additives are suitable too. You'll also need some moly for lubricating the cam.

Each of the five 351C and 351M/400 bearings is different. They get smaller from the front to the back. Bearing numbers and positions are printed on the box or on a separate slip like this. Note size difference between the front and rear cam bearings.

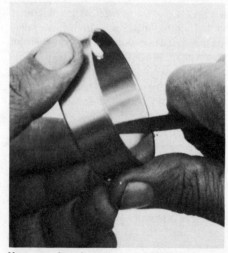
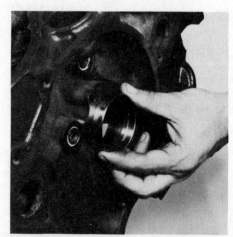

New cam bearings have square edges and possibly a slight burr which makes installing a cam difficult. Before installing bearings, chamfer their edges with a penknife or a bearing scraper as I'm doing here.

Holes in cam bearings are there for a purpose, to provide a passage from the oil galleries to the bearing journals. Front bearing insert has two holes, so be careful when lining up bearing for installation. Extra hole is for cam-journal-to-distributor-shaft oil gallery.

Installing cam bearings isn't much different than removing them, except considerable care must be taken to make sure they don't get damaged, the oil holes line up and they are square in their bores. Cone, pictured close up, centers drive-bar in bearing bores so bearing inserts are installed straight and square.

All but the front cam bearing can be eyeballed during installation. Front insert must be installed behind the front face of its bearing bore: 0.005—0.020 inch for 351Cs and 351M/400s or 0.040—0.060 inch for 429/460s. Check this using feeler gages and a straight edge or the thrust plate like this.

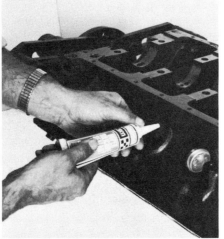

I use sealer to make sure the core plugs won't leak. Use a large-diameter punch to install core plugs. Whatever you do, don't hammer on the plug edge to install it. This distorts its edge and increases the likelihood of a leak.

Install the Bearings—Start with the front or rear bearing and work toward the center from the opposite end of the block. When you finish installing the center bearing, install the remaining two bearings by working from the opposite end. The reason is, the farther you are from the bearing—relative to its position in the block—the more accurate the installation tool can be lined up with the centerline of the cam-bearing bore regardless of whether it's self-centering or not. You'll have the bores at the opposite end of the block from the bearing you're installing to use as a reference to line up the bearing.

To install a bearing, select the one which fits the location you want to start with. Although obvious, most bearing manufacturers list bearing locations with each part number on the box. Cross-reference these numbers with those on the back-side of the bearing insert to confirm their locations. This is also a good check to make sure you have the right bearings. I once got a Buick V-6 bearing mixed in with four 351C bearings!

Begin installing the first bearing insert by slipping it over the mandrel. If you have the expansion-type mandrel, you'll have to expand it so it fits snugly in the ID of the bearing. With the solid type, select one that fits the bearing ID. Wrap some masking tape around the mandrel if it's a little loose. Also apply some oil to the mandrel.

With the pull or drive rod handy, position the bearing and mandrel over the bearing bore so the oil hole/s in the insert line up with those in the block—so they will line up after the bearing is installed.

Use sealer on the cam plug. It seals oil and is considerably more difficult to get to with the engine in place, so a leaky cam plug presents a big problem once the engine is installed. This engine is fully assembled because an engine stand was used to support it during its assembly making it impossible to install the cam plug. You'll have the same problem if you use a stand, so don't forget the cam plug.

ENGINE ASSEMBLY

If you have the threaded, or pull-type tool, locate the bearing and mandrel on the opposite side of the bearing web from which you'll be pulling. You'll put the drive-type mandrel on the same side from which you'll be hammering. Regardless of the type tool you're using, check the bearing immediately after you've started it in the bore to make sure it's going in straight, then finish installing it if it's OK. If not, straighten it up and finish the installation. Except for the front bearing, it will be positioned correctly when it looks centered in its web. The front bearing must be located more precisely.

Feeler Gage for the Front Bearing—The front edge of the number-one cam bearing must be located accurately *behind* the front face of the front bearing housing. When you think the bearing is close to being in position, check it with your feeler gages between a straight edge laid across the thrust-plate mounting surface and the front edge of the bearing insert. If the clearance is not within the specified tolerance range, move the bearing accordingly and check it again. Also, check to see if the bearing is installed squarely by gaging around the bearing with your feeler gages in three or four locations. If it's more than a few thousandths off, square it up. Number-one camshaft-bearing setback for 351Cs and 351M/400s is 0.005–0.020 inch. Use 0.040–0.060 inch for 429/460s.

Install All the Plugs—With the cam bearings in place, you can install the camshaft-bore plug. While you're in the plug-installation business, now's a good time to install the balance of the plugs that go in the block: 6 core plugs which fit in the sides of the block—3 per side—and 4 oil-gallery plugs, 2 in front and 2 in back.

Water-Jacket Plugs—Water-jacket, or core plugs are 1-1/2-inch diameter cup-type plugs. It's not essential to use sealer when installing these but I do anyway. Apply a small bead of sealer around the front edge of the plug or the outside edge of the hole and set the plug squarely over it so you're looking *into* the concave side of the plug. Use a hammer and the largest diameter punch you have—not over 1-3/8-inch diameter—to drive the plug into place. A punch 1/16-inch smaller than the ID of the plug is ideal. Don't use a punch less than 1/2-inch round because it may distort the plug, causing it to leak. Likewise, don't drive the plugs in by hitting them on their edges, otherwise the same thing will happen. Install them by driving a little at a time. Work around the inside edge of the plug, making sure it goes in

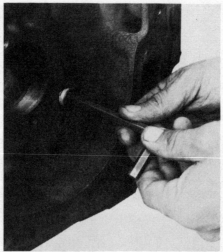

Short oil-gallery plugs go at front and long ones at rear. After applying sealer to threads, run them in and tighten them with an Allen wrench. A box-end wrench slipped over the Allen wrench gives additional leverage for tightening.

squarely. When the outer edge of the plug is just past the inside edge of the hole—say 1/32 inch—the plug is in far enough.

Cam Plug—The rear cam plug is also a cup-type—2-1/4 inches for 351Cs and 351M/400s and 2-1/2 inches for 429/460s. This plug must seal perfectly, particularly if you have a standard transmission, so use sealer. It will ensure that you won't have an oil leak that eventually appears as ugly spots on your driveway, or worse yet, as a well-oiled, slipping clutch.

Install the cam plug the same way you did the core plugs. Use some sealer and a large punch—again, not smaller than 1/2 inch. Drive the plug in just so its outside edge is past the edge of its hole and wipe the excess sealer off to make a neat job.

Oil-Gallery Plugs—To install the oil-gallery plugs, you'll need a 5/16-inch Allen wrench. If yours can be used with your socket set, great. A box-end wrench slipped over the end of the standard L-shape Allen wrench will give you the needed leverage. You undoubtedly found this out when removing the plugs. Fortunately you won't need as much leverage to install them. As with the camshaft plug, it is critical that the rear oil-gallery plugs seal. If the front ones leak slightly, it's not going to hurt anything because the oil ends up back in the oil pan. I coat front and rear plugs with sealer just the same. Also, because of the need for sealing at the rear plugs, Ford uses longer plugs at the back-end of the oil-gallery holes. Do the same. Coat the threads with sealer and run them in firmly. Wipe off excess sealer.

Replacing oil-filter adapter in 429/460 block. Torque it to 80 ft.lbs., being very careful not to round the corners of the hex.

Oil-Filter Adapter—If you removed the oil-filter-to-block adapter, reinstall it. You'll need a 1-1/4-inch socket. If you remember, the integral hex nut is short, so be careful when torquing it in place so the socket doesn't slip off and round the nut's corners. Torque the adapter to 80 ft. lbs.

CAMSHAFT INSTALLATION

How you install your camshaft and prepare it for its first few minutes of initial engine run-in—regardless of whether the cam is new or not—will establish how long it's going to live: 30 minutes, 30,000 miles or 100,000 miles. A not-too-uncommon result of an improperly installed camshaft is one or more lobes get wiped, or rounded off. The damage is not confined to the camshaft because metal particles from the cam lobe and lifter end up well distributed through an engine's oiling system—oil pump, filter, main bearings and everywhere oil is circulated. The oil filter traps most—but not all—of the debris. This means the cam and lifters have to be replaced, plus all the bearings because metal particles become imbedded in their soft aluminum-tin or copper-lead overlays. Having to do this immediately after a complete rebuild can make a grown man cry. So, be particularly careful during this part of the engine assembly and make your wallet smile.

Lubricate the Cam Lobes—Due to high contact pressure between camshaft lobes and their lifters and the possibility the lifters and lobes will not be receiving much lubrication during cranking for initial startup, the cam lobes must be lubricated with something to protect them during these first critical minutes. Here's where *molybdenum-disulfide*, commonly

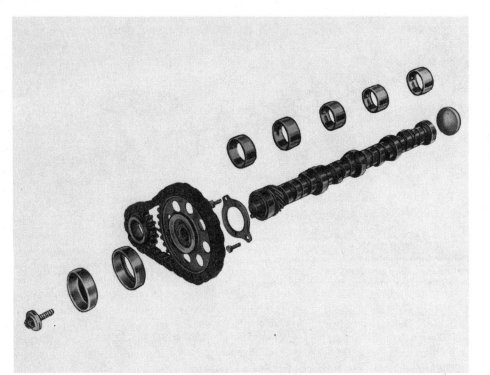

Camshaft and cam-drive-assembly components. Two-piece fuel-pump cam is used on later-model engines. Front, or center piece clamps tightly against cam sprocket, leaving outer ring free to rotate. Photo courtesy Ford.

If you are installing a new or reground camshaft, remove drive pin from old camshaft before turning it in as a core. Replacement cams usually don't include the pin. Remove yours by clamping on it with Vise-Grips and working it out.

known as *moly-disulfide*, comes in. Dow Corning's Molykote G-N paste is readily available in 2.8-oz. tubes, or if you're installing cams for a living get it in pint cans. Valvoline's moly-disulfide is Special Moly EP Grease.

Before installing the cam, wipe the cam bearings clean with lacquer thinner and a paper towel, then oil them well. Wipe the oil around the bearings with your fingertip. Do the same thing with the bearing journals. Apply moly grease evenly to all lobes and you're ready to install the cam. Be careful when handling the cam so you don't touch or get any dirt on the lobes. This is especially true if the cam is new because the black *phosphate coating* for oil retention also retains other materials which come in contact with the lobes.

Install the Cam—Temporarily install the sprocket onto your cam to provide a "handle." With your block on its rear face, feed the cam into place, being careful not to bang the lobes into the bearings. Lobes and bearings are pretty tough, but there is no point in needlessly damaging them when it is easily avoided. You can install the cam about halfway in the block by supporting it by its gear with one hand and by the center bearing journal with the other. To install it the rest of the way, reach inside the block with the hand you had on the bearing journal and carefully feed the cam into the engine until the journals line up with their bearings. Slide it into place.

Your camshaft should turn easily by hand. If it doesn't, the realization of why I tried to discourage replacing cam bearings will suddenly hit you between the eyes. Assuming the bearings were installed correctly, the problem will be that their IDs aren't aligned on the same centerline. Consequently, the cam is bound, because little to no clearance exists between some of the cam journals and their bearings. If you run into this problem, you'll have to depend on an experienced engine builder to correct the problem, otherwise you may find yourself in even more trouble. The factory align-bores the cam bearings to true them, whereas the engine builder will probably use the "bluing-and-scraping" technique.

Cam-Bearing Clearancing—Scraping bearings simply involves removing bearing material where clearance is needed. Determining areas needing clearanced is done by "bluing" the bearings—applying machinist's blue to the ID of the bearings—then installing the cam and rotating it a couple of revolutions. Blue is removed from the tight areas needing clearancing. After some material is removed in these areas with a bearing scraper, the bearings are reblued and the camshaft is reinstalled for rechecking. This is a long involved process because only small amounts of bearing material should be removed at a time to prevent over-clearancing. When the cam rotates freely and there are no apparent tight spots, the cam can be installed for good, but not until the block and cam are thoroughly cleaned and relubricated.

With the cam in place, install the thrust plate with two 1/4-20, grade-8 bolts. Just make sure the cast-in slots are facing the front cam-bearing journal. Holes in the thrust plate are staggered so you may have to rotate the plate until the holes line up with the threaded holes in the block. Torque the bolts 9—12 ft. lbs.

CRANKSHAFT INSTALLATION

Turn your attention to the crankshaft and its bearings and rear-main oil seal. You'll have to choose bearings based on the size of your main-bearing journals. The rear-main seal comes with your engine's gasket set.

Size the Bearings—Refer to your crankshaft inspection record to determine which size bearings to use—standard or 0.010, 0.020, 0.030, etc. undersize. For example, standard main-bearing-journal diameters are:

Engine	Main-Bearing-Journal Diameter (inch)
351C	2.7484—2.7492 (2.7488 nominal)
351M/400	2.9994—3.0002 (2.9998 nominal)
429/460	2.9994—3.0002 (2.9998 nominal)

If your main-bearing journals are in the

Rather than removing teeth marks from the pin, take advantage of them. They'll help secure the pin in the new cam. So install the pin end-for-end by driving it into place with a punch and hammer. When it bottoms, it's installed.

Before installing the camshaft, apply a coat of moly-disulphide (MoS_2) to the lobes. Don't skimp here. Put moly on the cam bearings too.

range shown in the chart, you'll obviously use standard bearings. However, if they mike some amount less, you'll need the same amount of undersize bearing as the difference from the standard nominal bearing size. Simply put if a 351C main-bearing journal mikes 2.7478 inches, the proper undersize bearing to use is 2.788 − 2.7478 = 0.010-inch undersize. So, to determine if and how much of an undersize bearing you need, subtract the diameter of your bearing journals from the standard nominal bearing-journal diameter.

There are two types of rear-main bearing seals you can install: the *rope* and the *split-lip* types. A rope seal will probably be included in your gasket set. It consists of two pieces of graphite-impregnated rope. The split-lip seal is made from neoprene, also in two pieces. If you want this seal instead of the rope type, you'll have to lay out another three bucks or so. I like the split-lip seal because a rope seal is more difficult to install. And, the rope seal creates more friction, as initially installed, meaning the torque required to turn the crank during assembly will be approximately 10 ft. lbs. more. However, the relative effectiveness of the two seals is the same. Also, the rope seal's initial friction drops off considerably after the first few minutes of engine operation.

Install the Main-Bearing Inserts—With the engine block upside down, wipe the bearing bores clean so there won't be any dirt trapped behind the bearing inserts. Wipe them clean using paper towels and a

Install cam by feeding it through one bearing web at a time. Challenge is to avoid bumping the bearing inserts so they don't get damaged. Control camshaft when installing it by using the loosely installed drive sprocket. Your cam should rotate easily once installed.

Here's how to install a camshaft if your crankshaft is already in place. Feed it in from the front while supporting it with the sprocket. Set the engine on its rear face to make it easier. You'll then have better control of the cam and gravity will be working for, not against you.

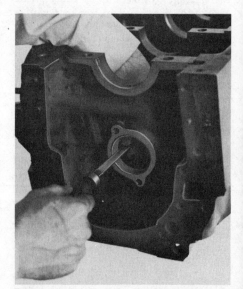

Bluing and scraping cam bearings—the thing I warned you about possibly having to do. I'm doing the front one here. Center three are tough to do. This is a job for an expert.

96 ENGINE ASSEMBLY

BEARING-CLEARANCE CHECKING ROTATING METHOD

The simplest way to confirm you've installed the correct size bearings for a given size bearing journal is, first rotate the journal in the bearing, or vice versa in the case of connecting rods, to check for insufficient clearance. If the rotation is free it's OK. To check for excessive clearance, try to rotate it 90° to the normal direction of rotation. If some play, or movement is felt, the clearance is too much. Let's look at how this applies to crank and rod bearings.

Crankshaft Bearings—Install the crankshaft using the methods outlined in this chapter. If you are using a rope-type oil seal, don't install it until after you've checked the bearings. With the crankshaft in place, the bearing caps installed and their bolts torqued to specification, rotate the crankshaft by hand. If it rotates freely, the bearing clearances are enough. To check for excessive bearing clearance, lift the crankshaft by its nose. Try to wiggle it up and down. If you don't feel any movement, or play, clearance is not excessive and you can proceed with your engine assembly.

Connecting-Rod Bearings—Connecting-rod bearings can be checked with the crankshaft out of the block. This is because the pistons cannot be installed in their bores for the check. For checking a rod bearing, install the rod on its bearing journal after you've oiled the journal or bearing. Torque the rod-bolt nuts to specification, then rotate the rod around on its journal. If it's free to rotate there is enough clearance. Now, try wiggling the rod sideways. If there's no movement the clearance is not excessive and the bearings are OK.

PLASTIGAGE METHOD

Plastigage is a strip of wax which, when installed between a bearing and its journal, flattens or squeezes out to a width inversely proportional to the clearance between the journal and the bearing. Plastigage comes in a paper sleeve with a printed scale for measuring the flattened Plastigage. It reads out directly in thousandths of an inch bearing-to-journal clearance. You bought green Plastigage to measure 0.001—0.003-inch clearance range.

To use Plastigage, cut a length of it to coincide with the bearing width you're checking. The bearing and journal must be free from oil because Plastigage is oil-soluble, and any oil will cause a false reading. When checking the crankshaft, *all* the other caps and bearings must be in place and tightened first. Lay the Plastigage on the top of the bearing journal or centered in the bearing, and carefully install the cap with the crankshaft laying in the block on its new bearings, but without its rear-main oil seal. Or, the piston and rod assembly must be positioned against its bearing journal. Torque the bolts or nuts to specification. Be careful not to rotate the bearing and journal relative to each other, otherwise the Plastigage will smear and you'll have to start over again. After you've finished torquing the cap, remove it and measure the bearing clearance by comparing the squeezed wax width with the printed scale on the Plastigage sleeve.

It's not necessary to check all the bearings unless you want to be particularly careful. One thing for sure, you can't be faulted for checking to be certain.

Line up staggered holes of the cam thrust plate, then install the two 1/4-inch bolts. Torque them 9—12 ft.lbs.

solvent such as lacquer thinner—both front and back. The same thing goes for the inserts too. The actual bearing surface will be coated with a white residue as it comes from the box. It's OK, just wipe the bearing enough so you're certain the working surface is clean.

429/460 crankshaft and related components. Note load-spreader ring just back of automatic-transmission flex plate. Some engines use it and some don't. If yours does, use it. Photo courtesy Ford.

ENGINE ASSEMBLY 97

Wide flanges of thrust-bearing inserts distinguish them from single-purpose radial bearings. Crankshaft thrust-bearing inserts locate in the center main-bearing web.

Sometimes bearing-box labels don't tell the truth. Check the bearings to confirm their size. They will be standard (STD) like this one, or so many thousandths undersized: -010, -020, etc.

Using Plastigage to check bearing clearance. Lay a strip of Plastigage the full length of the bearing journal (arrow), then install bearing and cap with bolts torqued to specification. Make sure journal is free from oil, and the other bearing caps are torqued *first*. After final torquing, carefully remove cap and bearing and check width of *squeezed* Plastigage—it corresponds to the bearing clearance. This Plastigage stuck to the bearing. Sometimes it stays on the bearing journal.

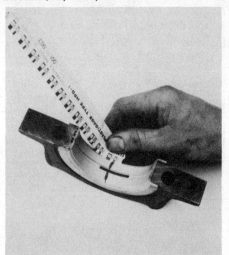

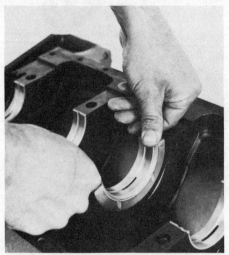

Before installing bearing inserts, wipe bearing bores and the back of the inserts clean. Inserts having grooves and holes go in block and plain ones go in caps. Flanged thrust-bearing inserts require more force to seat them.

Pointed pin in rear main-bearing-cap seal groove prevents rope seal from rotating—don't remove it if you're using a rope seal. If a split-lip seal is to be used, remove the pin from the back side of the cap with a 3/32-inch or smaller punch. Fill hole with silicone sealer to prevent an oil leak.

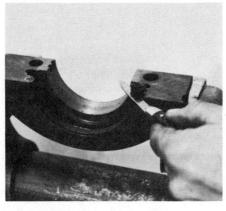

Large socket and brass hammer being used for installing a rope seal in the bearing cap by laying the cap in vise jaws. Don't clamp on the cap. Work around seal until it is all the way in its groove, then trim ends flush with cap or block. Install seal in block the same way.

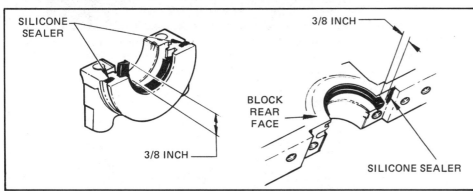
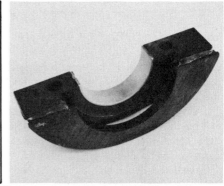

Drawing shows how to position a split-lip-type seal. Each half should project approximately 3/8-inch out of its groove in the bearing cap and block, but on opposite sides so the two will engage when the cap is installed. Apply a *small amount* of silicone sealer just prior to installing the cap. Here's a cap with its bearing insert, seal and sealer ready to be installed.

Each bearing half has a bent-down tab at one end. The top bearing halves which go into the block are grooved. Their tabs are located off-center whereas the bottom halves are not grooved and their tabs are centered. The locating tabs help prevent the bearings from being installed wrong.

Install the bearings in the block by first putting the tab of each bearing in the notch at the edge of its bore. Hold the bearing flush with the edge of the bearing bore with a finger or thumb while you force the bearing into place by pushing down on the opposite end with your other thumb. All the bearing halves will go in with little effort except for the center one. It also serves as a thrust bearing, and will require more force because the thrust flanges fit tightly around the bearing web.

Before installing the top bearing halves in the caps, *lightly file* the cap-to-block mating surfaces to ensure the caps will fit in their registers. To do this, you'll need a large flat, fine-tooth file. Lay the file on your bench and stand a cap up on the file. Lightly run the cap over the file a few times while holding the cap square against the file. Be careful not to remove any material from the cap except for nicks or burrs—small projections raised above the normal mating surface. After doing this, install the non-grooved inserts into the caps.

Check Bearings-to-Journal Clearance—You have already determined the correct size bearings for your main bearings by measuring the main journals with a micrometer. Due to the relatively close accuracy of mikes and the close tolerances to which bearings are manufactured, direct clearance checks are unnecessary, assuming the right bearings got in the right box. However, assuming can cause considerable trouble and checking merely takes time.

Rear-Main-Bearing Seal—If you have a split-lip rear-main-bearing seal, you can install it now if you aren't clearance checking with Plastigage. If you have the rope type, install it later if you are checking main-bearing clearances now using either method. With this in mind, let's get on with installing both types of seals.

Split-Lip-Seal Installation—If your engine was originally equipped with a rope seal—chances are it was as it came from the factory—a sharp pin is centered in the seal groove in the rear-main cap.

Remove the pin with a small punch, driving it out *from the back-side in*. If you've already installed a bearing insert in the cap, remove it so it won't get damaged. Reinstall it after you have the seal in place. Place the cap over the end of a piece of wood so the load taken by the cap is directly below the pin as you drive it out.

With the rope-seal pin out of the cap, fill the remaining hole with silicone sealer, install the seal and reinstall the bearing insert. Lightly coat the seal halves with oil, and while installing them, don't let oil get on the block and cap mating surfaces. If it does, wipe them clean with lacquer thinner. The seal lip *must point toward the front of the engine* when installed. Also, don't line up the ends of the seal halves with the cap and block mating surfaces. Rotate the seal halves in the block and cap so one end projects approximately 3/8 inch up from the block and cap mating surfaces and so they fit together with the cap installed.

Rope-Seal Installation—Put half of the rope seal in the block groove, laying it edgeways in the groove. Force it in with your thumb, leaving both ends of the seal extending above the block and cap

Prior to installing the crankshaft, lightly wipe main bearings and crank journals with a clean paper towel, then oil them. Spread oil evenly with a finger tip—lubricate seal too.

Although heavy, carefully lower crankshaft squarely onto its bearings. Watch your fingers at the rear seal. You'll need a finger in the end of the crankshaft to support it.

mating surfaces. With the seal in place, work it into the groove with a cylinder— I use a 1-1/2-inch socket—and a hammer to tap on it as you roll the cylinder back and forth. You don't have to pound it in, just use a light tapping motion. When the seal has filled the groove, trim the excess ends off flush with the block using a sharp knife or a single-edge razor blade.

Install the other seal half in the bearing cap. It'll be more difficult just because the cap will be clumsy to handle while you're installing the seal. It's a whole lot easier if you cradle the cap in something like a vise. *Don't clamp the cap in a vise.* Just open the jaws far enough to permit the cap to sit squarely. Install the seal as you did in the block. When the seal is bottomed in its groove, trim the ends off flush.

Get Your Crankshaft Out of Storage— With the main bearings and rear-main seal in place you are now ready to drop your crankshaft in the block. If you are going to check bearing clearances don't install the crank seals until you've checked the clearances. Anyway, you can now reintroduce your crankshaft to daylight. Use solvent and paper towels to clean all of the bearing journals of their protective oil coating. They will have collected some dust and dirt particles by now and spray-type oil is not suitable for engine assembly anyway.

If you are installing your crankshaft to stay, check its oil holes again for dirt. This is your last chance to catch any contaminants that will otherwise be pumped to the bearings the instant you crank your engine. Oil the bearings in the block as well as those in the caps. Also spread some oil on the seal. Lower the crank carefully into the block.

Now you're ready for the caps. With their bolts lightly oiled—threads and under their heads— and loosely installed

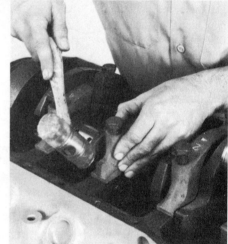

As you did with block bearing inserts, wipe the cap inserts clean and oil them. Loosely fit the caps to the block in the right *position and direction*. After oiling threads, run the bolts part way into the block, then tap caps on their sides like this to seat them in their block registers.

Start by torquing all main caps except the center one. I'm starting with the rear cap so I can check rope-seal drag. It was 14 ft.lbs. If you're using a rope-type rear-main seal and you've made all the necessary crankshaft checks, don't get alarmed when you find your crankshaft is a bit hard to turn.

in the caps, position them on their bearing journals using the cast-in numbers and arrows for location and direction. The number must coincide with the bearing-journal number and the arrow must point to the front of the engine. The rear-main cap won't have a number or arrow. Its position and direction are obvious.

Seal the Rear-Main Cap—For final crankshaft installation, the rear-main-bearing cap and cylinder block parting line must be sealed, otherwise it will leak. Run a small bead of silicone sealer—about 1/16-inch wide—in the corners of the block's rear-main register from the back edge of the block to even with the front edge of the crankshaft oil-slinger groove. Also, apply another 1/16-inch-wide strip of sealer to the cap in from both edges of the cap and in line with the seal. Stop about 1/8-inch short of the seal. Refer to the sketch on page 99 to see how to apply sealer. Oil the seal and bearing before installing the cap.

Don't Tighten the Main-Bearing Caps Yet—Before threading the bearing-cap bolts in the block, the caps should be located in their registers. If the bolts are tightened in an attempt to draw the caps into place, you may damage some caps and the block. The edges of the bearing register or caps can be broken off, even though the block is chamfered. Locate one edge of each cap against its register; then tap the cap on its opposite side at about a 45° angle while pushing down on the cap. It should snap into place. After the caps are seated you can run the bolts down, but don't tighten them yet.

If you have the rope-type seal, now is the time to see how much torque is required to turn your crankshaft against its friction. Tighten the rear-main cap only. Do each bolt progressively. Don't run one bolt down, snug it up and torque it, then go to the other one. The right way to do this is to run them both down, snug one, then snug the other and torque both to specification: 1/2-13 main-bearing bolts get torqued 95–105 ft. lbs., 7/16-14 are torqued 70–80 ft. lbs. and 3/8-16 to 35–45 ft. lbs. All 351C, 351M/400 and 429/460 engines use 1/2-13 main-bearing bolts. Performance versions of these engines use four-bolt main-bearing caps with smaller bolts on the outboard side of the caps. The 351C-CJ, Boss and HO engines use an additional 3/8-16 bolt whereas the 429 CJ, SCJ and Police engines use a larger 7/16-14 bolt.

Now, install the damper attaching bolt and washer so you can turn the crank. Expect 10–15 ft. lbs. to turn it if you

Seating thrust bearing insert. Before tightening center-main-bearing-cap bolts, seat its bearing. Wiggle crankshaft back-and-forth, then hold it forward while tightening bolts.

Two methods of checking crankshaft end-play: dial indicator and feeler gages. Dial-indicator is the most accurate because it checks actual crank movement. Crank is forced in one direction, indicator is zeroed, then crank is forced back in the other direction. End-play is read directly from indicator. It is safe to assume that crankshaft end-play approximates the clearance between a thrust face and its bearing when the clearance on the opposite side is eliminated by moving the crank.

used a rope seal. At least you'll know what's requiring the high effort to turn your crankshaft during the rest of the engine assembly.

Seat the Thrust Bearing—Using the same technique you used for torquing the rear-main-cap bolts, secure the others except for the center cap—the one with the thrust bearing. Tighten its bolts finger-tight, then line up the bearing thrust faces by forcing the crankshaft back and forth in its bearings. Use a couple of big screwdrivers to pry between another bearing cap and a crankshaft throw or counterweight. After doing this a few times, pry the crankshaft forward. Hold it in this position while you torque the two bolts which secure the thrust-bearing cap. Check crankshaft-turning torque again. You should find negligible increase.

Tools for installing piston-and-connecting-rod assemblies. Bearing-journal protectors for the rod bolts, ring compressor, hammer for pushing the piston down into its bore and oil in a squirt can and tomato can for oiling piston and rings.

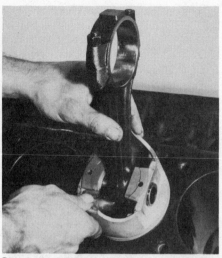 

Check ring end-gaps before installing rings on the pistons. First, square each ring in its bore like I'm doing with this piston. Use your feeler gages to check end gap. This is a precautionary operation as standard rings are pre-gapped. What you are doing is confirming that you have correctly gapped rings. Gaps should be about 0.016-inch.

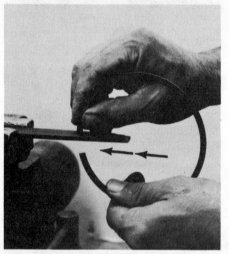 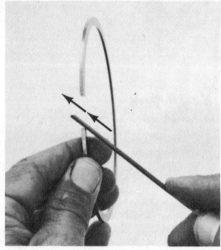

Increasing ring end gap by filing one end of ring with a fine-toothed file. With file firmly clamped, move ring against file *in direction of arrow*. Lightly touch corners of end you filed to remove any burrs.

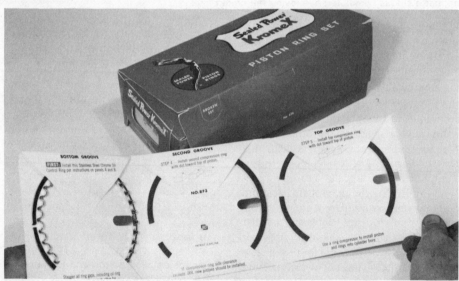

Pistons rings come neatly packaged and include explicit instructions as to how to install them. Each individual set of rings has its own package as shown here. They are arranged in the same order as they are installed; oil-ring assembly, bottom compression ring and top compression ring.

Crankshaft End-Play—Crankshaft end-play should be 0.004–0.010 inch with a 0.012-inch maximum. To check end-play accurately, a dial indicator is needed. However, if you don't have one, feeler gages will do an adequate job. To use the feeler gages; pry the crankshaft in one direction, then measure between the crank's thrust face and bearing. However, if you have a dial indicator, mount it with its plunger parallel to the crankshaft centerline and with the plunger tip firmly against a flat surface on the crankshaft's nose or flywheel-mounting flange. Pry the crank away from the indicator, then zero the dial gage. Now, pry the crank in the other direction to take up its end-play. Read total end-play directly. Repeat this a few times for verification.

If, by rare chance, end-play is under the 0.004-inch minimum, the thrust-bearing flanges will have to be thinned, meaning the crankshaft must be removed! Thin the thrust bearing's front flange with 320-grit sandpaper on a *flat surface*, *lapping* away bearing material to bring end-play within tolerance. Before doing any sanding, use a micrometer to measure the width across the bearing flanges. Check it at both ends and in the middle. Do both bearing halves and record the initial results. While sanding, hold the bearing flat on its thrust flange and avoid any rocking motion as you move it back and forth. Periodically measure to make sure you are removing material evenly and not excessively. When you've matched both bearing halves, clean and reinstall them and the crankshaft following the correct procedure. Check end-play again to confirm what you did produced the right effect.

If the end-play exceeds 0.012 inch, trade your crankshaft in on a crank kit and let the reconditioner worry about the worn one. Too much end clearance is highly unlikely considering you've inspected your crankshaft and declared it sound, or had it reconditioned or replaced. However, unlikely or not, end-play is one of those items which must be checked regardless of the consequences. Like the man says, "pay now or pay later." What you'll have to pay now will be a fraction of the cost later in both time and money.

PISTON AND CONNECTING-ROD INSTALLATION

If you've yet to assemble your pistons and connecting rods, refer back to page 70 for how it's done. However, if they are assembled, you can prepare them for installation. The equipment you'll need

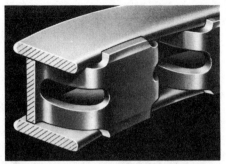
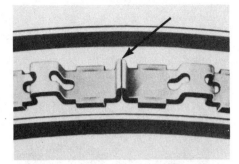

Section of an oil-ring assembly. Left drawing shows how the expander-spacer fits behind the two rails to spring-load them against the cylinder bore. Ends of a correctly installed expander-spacer must butt (arrow), as shown in right drawing. Courtesy Sealed Power Corporation.

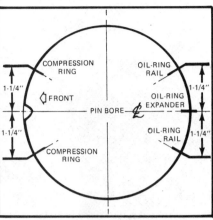

Locate ring end gaps like this prior to compressing them for installing the piston-and-rod assemblies into your engine.

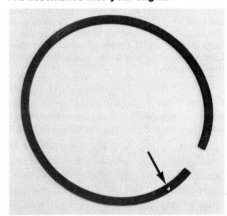

Pip mark (arrow). When installing a compression ring, make sure mark is toward piston top.

includes a small flat file, a ring compressor, hammer, a large tomato can containing about a half quart of clean motor oil, an oil can and some rod-bolt bearing-journal protectors. You should consider a ring expander too, even though it's possible to install the rings without one. The expander ensures the pistons won't get scratched or the rings bent or broken during the installation. If you can't readily pick up a couple of bearing-journal protectors, two 2-inch lengths of 3/8-inch rubber hose will work.

Gap Your Rings—Piston rings are manufactured to close tolerances to fit specific bore sizes, however, you should check their end gaps just the same. Just as with bearings, the wrong rings can find their way into the right boxes. Once they're on the pistons and in their bores, you'll never know until it's too late. Also, some high-performance piston rings must be gapped.

Both compression rings should have a 0.010—0.020-inch end gap, or more precisely, 0.004-inch gap for each inch of bore diameter. For example, compression rings for a 4-inch bore should be gapped at 0.004 x 4 = 0.016 inch. However, if your ring gaps are in the 0.010—0.020-inch range, they'll be all right. Gap range for the two oil-ring side rails is much wider at 0.015—0.055 inch.

To check a piston ring's end gap, install it in a cylinder bore. Carefully fit the ring in the bore and square it with an old piston or a tin can. There's no need to push the ring down the bore more than a half-inch unless your block hasn't been rebored. In this case you should push the ring at least 4 inches down the bore. The reason is taper in a cylinder causes a ring to close near the bottom of the stroke and *minimum* end-gap-clearance should be checked to prevent the ends of a ring from butting when the engine is at maximum operating temperature. If any gaps are too large, return the complete set of rings for replacement. If a gap is too small, it can be corrected by filing or grinding the end of the ring.

When filing or grinding a piston ring, the first rule is to file or grind the end of the ring with the motion *from its outside edge toward its inside edge*. This is particularly important with moly rings. The moly filling on the outside periphery will chip or peel off from the outside edge. When filing, don't hold the ring *and* file in your hands, clamp the file in a vise or onto a bench with a C-clamp or Vise-Grip pliers and move the ring against the file. You'll have better control using this method. For grinding a ring, use a thin grinding disc, one that will fit between the two ends. Just make certain the disc rotation grinds the end from the outside in. Hold the ring securely so you don't break it as you grind it.

Break the sharp edges from the ends of any rings you filed or ground. Use a *very-fine-tooth* file or 400-grit sandpaper. The inside edge will have the biggest burr, the outside edge the least—if any at all—and the sides will be somewhere in between. Just touch the edges with your file or sandpaper. Remove only enough material so you can't feel any sharp edges with your finger tips.

When you've finished one complete ring set, keep it with the piston and rod assembly that's to go in the bore you used for checking and fitting. Repeat this procedure until you've checked and fitted all the rings. It's best to keep the rings in the envelopes they came in. The ring manufacturer usually organizes the rings in sets with its rings in order—top compression ring, second compression, then the oil-ring assembly. Be careful not to get the rings mixed up, particularly the compression rings. The differences between them are not readily apparent, so it's best to keep the rings in order to avoid installing them on the piston in the wrong order.

Install the Rings on the Pistons—With the rings gapped, you're ready to install the rings on their pistons. This can be done without any special tools, but I recommend using a *ring expander*. It greatly reduces the possibility of ring or piston damage during installation. An expander is cheaper than another set of rings.

So you won't have to horse around trying to hold a piston and rod while installing rings, lightly clamp each rod and piston assembly in a vise. This will leave your hands free to concentrate on installing the rings. Clamp the rod between two small blocks of wood with the rod far enough down in the vise so the piston can't rock back-and-forth. Don't clamp on the piston. If you don't have a vise, make yourself comfortable by sitting down in a chair and clamp the rod and piston between your knees. Installing rings this way is about as easy as using a vise.

Oil Rings—When installing rings on a piston, start with the oil ring and work up. You won't need the ring-expanding

ENGINE ASSEMBLY 103

Oil rings are easy, but tricky to install. After the expander-spacer, top rail goes on next. Now's the time for checking the expander-spacer ends. They shouldn't overlap. If they do, remove the rail, correct the problem and reinstall it. When installing a rail, hold its free end so it doesn't scratch the piston as you bring it down over the piston and into its groove. Position the rail and gaps now.

Two ways of installing a compression ring—with thumbs or with an expander. If you elect not to use a ring expander, be careful. It's easy to bend or break a ring. If this happens you could've saved money by purchasing and using an expander. Regardless of how you elect to do it, remember the pip marks go up.

tool for the oil-ring components—the expander-spacer and the two side rails. However, you can save yourself some time later on by installing the three rings which make up the oil-ring assembly so their end gaps are in the right relationship to one another rather than waiting until it's time to install the piston and rod in its bore. The idea is to *stagger* the gaps so they don't coincide with each other. If this were to happen the oil ring could pass excessive amounts of oil.

The piston front is shown by a notch or arrow stamped in its dome. A cast piston has the notch, and a forged one has an arrow. The drawing on page 103 shows how to position the end gaps for all rings. Install the expander-spacer with its gap directly at the front of the piston. Its job is to locate the two rails in the oil-ring groove. Be careful when installing the expander-spacer. Make sure its ends butt and don't overlap. You really have to work at keeping this from happening with some expander-spacer designs.

Now you're ready for the rails. Insert one end in the oil-ring groove *on top* of the expander-spacer. Hold the free end of the ring with one hand while running the thumb of the other hand around on top of it to slip it into its groove. Don't let the free end of the ring dig into the piston as you bring it down over the piston and into its groove. Install the other rail below the expander-spacer using the same method. Position the rail gaps as shown in the drawing. Double-check the expander-spacer to make sure its ends aren't overlapping.

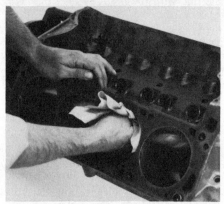

Wiping cylinder bores to remove any dust or dirt which may have accumulated. Don't use a cloth towel or rag, use a paper towel.

Compression Rings—Compression rings are much simpler in design, but more difficult to install than oil rings because they are stiff. They are easily damaged, and can damage the piston during installation if done incorrectly. Keep your eyes open when installing these guys. Look for the "pip" mark—a little dot or indentation—which indicates the top side of a compression ring. The *pip mark must be up* when the ring is installed, otherwise the *twist* will be in the wrong direction, causing excessive blowby and ring and ring-groove side-wear.

Twist describes how a ring bends, or the shape it takes in its groove. Twist is used for oil control and for sealing combustion pressures. Compression rings with no twist have no pip marks. If you get a set like this, although it's unlikely, you can install them either way. But inspect the rings to be absolutely sure. Also, read the instructions which accompanied the rings and follow them *exactly*.

The difficult thing about installing a compression ring is spreading it so it'll fit over the piston while not twisting or breaking the ring or gouging the piston with the ring ends. This is why I recommend using a ring expander.

There are different styles of ring expanders. The top-of-the-line type has a circular channel for the ring to lay in while it is expanded. The less-exotic expander doesn't have this channel. You have to use one hand for expanding the ring while controlling the ring with the other. Regardless of the type you have, don't spread the ring any more than necessary to get it over the piston. Install the second (lower) compression ring first. Make sure the pip mark is up, expand the ring while making sure it doesn't twist and install it over the piston into the second groove. Do exactly the same with the top ring, installing it in the top groove. End-gaps on these two rings should be placed as shown in the drawing. Check them for position *immediately before* installing the piston in the engine.

You may decide to install your compression rings without the aid of a ring expander, so I'll tell you how to do it right. Wrap the ends of your thumbs with tape. The sharp ends really dig in when you're spreading the ring. Position the second compression ring over the piston with the pip up. Rotate the ring over the edge of the piston so its ends are lower on the piston than the rest of the ring. With your thumbs on the ends of the ring so you can spread it and your fingers to the side for controlling it, spread the ring and rotate it over the piston with the

Piston-and-rod assemblies lined up as they would be installed; piston arrows and notches to the front and rod numbers toward their cylinder banks.

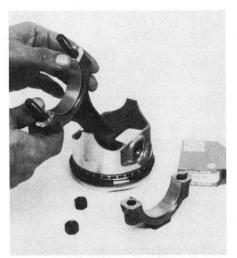

Bearing inserts complete the piston-and-rod assemblies. Install the bolt sleeves, oil the bearings, then immerse the piston in oil over the rings and pin. It's now ready for the ring compressor.

end gap over the groove. When the ring lines up with the groove all the way around, release it into its groove. When your thumbs revive, install the top ring using the same method. It'll be easier because it won't have to go down over the piston as far.

PISTON AND CONNECTING-ROD INSTALLATION

To install a piston-and-connecting-rod assembly into its engine, you must have a ring compressor, two bearing-journal protectors for the rod bolts, an oil can, something to push the piston into its bore—a hammer will do—and a large tomato can containing some motor oil. You don't have to have the last item, but it's the most convenient way I know to pre-lube your rings and pistons.

Get Everything Ready—Just like the other facets of engine building, you have to be organized when installing connecting rods and pistons. All tools must be within reach, everything must be clean and the engine positioned so you'll be able to insert each piston-and-rod assembly into its bore while guiding the rod into engagement with its crank journal.

Before starting the actual rod-and-piston installation, clean your engine's cylinder bores. Use paper towels, not rags. Look your pistons and rods over too. Regardless of how clean the rest of the assembly appears, wipe off the bearing inserts as well as their bores in the rods and caps. If you're doing the engine assembly on a bench, roll the block upside-down so one of the deck surfaces hangs over the edge of the bench. The engine will be resting on the other deck surface. This will give you clear access to the top and bottom of the bores and to the crankshaft-bearing journals. Have the pistons close by and organized according to their positions in the engine. This way you won't have to hunt for each piston even though it gets easier as you go along. With all the tools ready, you should be ready to slip the rods and pistons in their holes.

Crankshaft Throws at BDC—If you've removed the crankshaft damper bolt and washer, replace them. You'll need them so you can turn the crank during the rod-and-piston installation process. The throw or connecting-rod journal must be lined up at BDC with the bore that's to receive a piston and rod. This is so you'll have room between the bottom of the cylinder and the crank journal to guide the rod's big end into engagement with its journal as you slide the piston into the bore.

Once you have the crankshaft in position, remove the bearing cap from the rod you're going to install. Be careful not to knock the bearing inserts loose if you installed them earlier. Above all, don't mix them with other inserts. Avoid this by only removing one rod cap at a time. Slip the journal protectors over the rod bolts. Liberally oil the piston rings, the skirt and the wrist pin. Here's where the tomato can with oil in it comes in handy. Just immerse the top of the piston in the oil over the wrist pin, then spread oil over the piston skirt. While you're at it, spread oil onto the bearing inserts with a finger tip.

Ring compressors vary in design, so follow the directions that accompanied yours. If you don't have the directions, here's how to use one. There are two basic styles. Most common and least expensive operates with an Allen wrench. The compressed position is held by a ratchet and released by a lever on the side of the compressor sleeve. The second type uses a plier arrangement. A ratchet on the plier handle holds the rings in the compressed position. This type is available from Sealed Power Corporation with sleeves or bands to accommodate 2-7/8–4-3/8-inch piston diameters.

To use a ring compressor, fit it loosely around the piston with the long side of the sleeve pointing up in relation to the piston. With the clamping portion of the compressor centered on the three rings, compress the rings while wiggling the compressor slightly. This helps to compress the rings fully.

Using the notch or arrow on the piston dome or the rod number as reference for positioning the assembly, insert the piston and rod in its bore while being careful not to let the connecting rod bang against the cylinder wall. The lower portions of the piston skirt should project out of the bottom of the compressor sleeve. Insert the piston in the bore and push the compressor against the block to square it with the block deck surface. Holding onto your hammer by its head, use the handle to push or tap lightly on the head of the piston to start it into the bore. If the piston hangs up before it's all the way in, **STOP**. Don't try to force the piston in the rest of the way. What's happened is a ring popped out of the bottom of the compressor before entering the bore, consequently if you try forcing

With the crank positioned at BDC and the rings compressed, guide rod and piston into its bore. Make sure piston notch or arrow is pointing to the front. Push or lightly tap piston into its bore. If you feel any slight hangup, STOP!! Pull the piston out and start over after recompressing the rings. When ring compressor releases, push piston down into bore while guiding connecting rod into engagement with bearing journal.

Install rod cap and nuts now, but don't torque nuts until after you have all pistons and rods installed.

the piston any farther you'll probably break a ring and perhaps a ring land, resulting in a junked piston. Release the compressor and start over. It's no big loss, just a couple of minutes lost, as opposed to a piston and ring. When the last ring leaves the compressor and enters its bore, the compressor will fall away from the piston. Set it aside and finish installing the piston and rod.

Using one hand to push on the piston and the other to guide the rod so it engages with its rod journal, guide the rod bolts so they straddle the bearing journal, then tap lightly on the piston dome with your hammer handle until the rod bearing is firmly seated against the journal. Remove the sleeves from the rod bolts and install the bearing cap. Make sure its number coincides with the one on the rod and is on the same side. Install the attaching nuts after oiling the bolt threads and the nut bearing surface on the rod cap. Torque the nuts 40–45 ft.lbs., 45–50 ft.lbs. for 351C Boss and HO.

Now that the first rod and piston is installed, turn the crankshaft so its throw lines up at BDC for installing the next rod and piston. Finish with this cylinder bank, then roll the engine over so it rests on the other deck surface and do the other bank. With all your pistons and connecting rods installed you can step back and take a good look at your engine—it's beginning to look like one.

TIMING CHAIN AND SPROCKETS

Before you can install the timing chain and its sprockets, you'll have to rotate the crankshaft and camshaft to their cylinder-1 firing position. Bring piston 1 to the top of its bore and the keyway in the crank nose will be pointing up in relation to the block. Position it slightly left of straight up, looking at the keyway from the front of the block. You'll see why later. Positioning the cam won't be as apparent. Locate it so its sprocket drive pin is directly in line with the crankshaft.

Don't Forget the Key—351C and 351M/400 engines use a single 3/16-inch x 1-3/4-inch key to locate the timing sprocket and crankshaft damper. Two keys are used in the 429/460s, both of the *Woodruff* design. Woodruff keys are radiused on their backsides as viewed from the side. A 3/16-inch x 1-3/4-inch key is used at the sprocket and the other, a 1/4-inch x 7/8-inch key is used at the crankshaft damper.

To install a key, feel the edge of the keyway slot in the crankshaft slot. If it has a burr or raised edge, remove it with a file. Check the key too and smooth off any burrs with your file. Install the key by lightly tapping it into place with a punch.

Install the Crankshaft Sprocket First—If the damper bolt and washer are still in your crank, remove them so you can install the crankshaft sprocket. Don't slide it all the way on, just far enough so its keyway slot begins to engage the key. The timing mark on the sprocket should be visible and it should point toward the camshaft.

Before installing the cam sprocket, find its timing mark on the front face between two teeth, in line with the drive-pin hole. This mark must point down, lining up with the crank sprocket when cylinder 1 is on TDC. Using the mark as a reference, hold the sprocket so the timing marks on the two sprockets line up—directly opposite one another. While lifting up on the cam sprocket to keep the chain tight, slide the crank sprocket back simultaneously with the cam sprocket and chain until the cam sprocket engages the cam nose and drive pin. You won't be able to slide it on all the way because the drive-pin hole in the sprocket should be to the right of the cam drive pin. This is due to the position you left the crank in—to the left, or counter-clockwise. To bring the drive pin and its hole in line with each other, reinstall the damper bolt and washer in the end of the crank and move the crank clockwise while pushing on the cam sprocket. When it's lined up, you'll feel it engage the drive pin. You can now slide both sprockets all the way on. Before going any farther, make sure both sprocket timing marks point straight at each other. If they aren't, you'll have to back both sprockets off until the cam sprocket is loose and jump the chain a tooth in the right direction and reinstall the chain and sprockets and recheck them.

106   ENGINE ASSEMBLY

A simple check to make sure pistons are in the right holes and in the right direction—rod numbers pointing toward their cylinder bank and piston arrows or notches pointing forward. These pistons are numbered because they were fitted to the bores to obtain consistent piston-to-bore clearances. Even the most accurate pistons or bores are not exactly the same diameter, but don't think you have to do this because fitting pistons is not a standard rebuilding procedure.

Checking connecting-rod side clearance. It should be 0.010—0.020 inch with a maximum allowable clearance of 0.023 inch.

Align crank and cam like this when installing timing chain and sprockets. Crankshaft key is slightly left of straight-up.

With crankshaft sprocket and damper bolt and washer on crank end, fit cam sprocket and chain to crank sprocket so timing marks line up. Slide sprocket-and-chain assembly back so cam sprocket engages camshaft nose. It's going to be tight if you are installing a new chain and sprockets. Fully engage cam sprocket with its drive pin by turning crankshaft clockwise while pushing back on cam sprocket.

Install the Fuel-Pump Cam—Secure the camshaft sprocket, making sure you install the fuel-pump cam, or eccentric between the sprocket and the attaching bolt and washer. There are two basic types of cams, the one-piece and the two-piece. One-piece cams were used on the early 429s until they changed to the two-piece design during the 1971 model year. 429CJ, SCJ and Police engines continued to use the one-piece cam. The Police version of the 460 as well as 429/460s installed in F150 or larger trucks also use the one-piece design. All other fuel-pump cams are the two-piece design.

The difference between these two fuel-pump cams is the fuel-pump actuating arm *slides* against the one-piece cam much in the same way a lifter slides against its cam lobe. The outer ring of the two-piece cam rotates on its inner piece, reducing sliding action between the fuel-pump arm and the cam. Sliding action is not harmful to the cam or arm. The two-piece type is less costly to manufacture, which is why the change was made.

As opposed to most one-piece fuel-pump cams for other Ford engines, the 429/460 types doesn't depend on the camshaft drive pin to locate or drive it. It depends on the clamping force of the camshaft-sprocket bolt. Consequently it only has one hole in it. This doesn't present any particular problem unless the bolt loosens. If this happens, the cam rotates on the bolt rather than against the fuel-pump arm and no fuel is pumped to the carburetor. In a way this is a good feature because it brings attention to a bigger problem—the loose bolt.

Unlike the one-piece cam, the two-piece locates to the cam sprocket with a

ENGINE ASSEMBLY

bent-down tab on the back of the inner part of the cam. It fits loosely in the cam drive-pin hole.

Install the fuel-pump cam using the 3/8-16 x 2-inch or 1-1/2-inch-long, grade-8 bolt—depending on whether you have a 351C and 351M/400 or a 429/460. A dab of Loctite on the bolt threads is good insurance against the bolt loosening. With a washer behind the bolt, attach the cam to the camshaft nose and timing-chain sprocket and torque it to 40-45 ft. lbs.

Slide the oil slinger over the end of the crank extension and you're ready to install the timing-chain cover. Don't worry about the loose slinger. When the crankshaft damper is installed, it will sandwich the slinger (and spacer sleeve in the case of the 429/460) between it and the crank sprocket.

TIMING-CHAIN COVER AND CRANKSHAFT DAMPER

If you haven't cleaned the front engine cover yet, do it now. Remove all the dirt, grease and old gasket material. Use a punch and a hammer to remove the crankshaft oil seal. Carefully knock the seal out the back of the cover by placing the punch behind the inside lip and gradually work around the seal until it falls out. Don't try to knock the seal out with one whack or you may break the die-cast aluminum cover of your 429/460. With the old seal out of the way, clean the seal bore in preparation for receiving the new seal.

Install the New Seal—You'll find a new oil seal in your gasket set. Coat the outside periphery of its shell with sealer. How you install it depends on whether you have the 351C, 351M/400 sheet-metal cover, or the 429/460 aluminum cover. The seal is best installed in the sheet-metal cover by clamping the seal and cover between the jaws of a vise and using the vise like a press. This forces the seal straight in without it cocking. If you don't have a vise, you can install the seal with the cover on a bench. Use a socket or piece of pipe slightly smaller than the seal OD—2-1/2-inch diameter is about right—which will seat against the seal edge. Doing the installation this way is frustrating because the seal will cock, but it can be done with some care and patience.

An aluminum cover seal can be installed similarly, but the seal must go in from back-to-front rather than from the front side of the cover. To install the seal on a bench, support the cover immediately in front of the seal bore, then use a 2-1/2-inch OD socket or piece of pipe to drive the seal in as I just described for the sheet-

Complete cam-drive assembly by installing fuel-pump cam. You may have one of three cam types depending on your engine. Locate the cam over the sprocket and secure it with the attaching bolt and washer. Small hole in one-piece cam locates over camshaft drive pin. In case you have a 429/460 and are wondering, its fuel-pump cam doesn't have the small hole. The cam depends on bolt clamping force to prevent it from rotating relative to the cam. Torque the bolt 40—45 ft.lbs. Slip crankshaft oil slinger back against crank sprocket and you're ready for the front cover.

Rather than using camshaft drive pin, two-piece fuel-pump cam locates to cam sprocket with bent-down tab projecting into sprocket drive-pin hole. Outer ring turns on inner cup.

If you have 351C or 351M/400, replace its front cover if it's rusted like this one. Rusting through results in water being pumped into the crankcase. Using anti-freeze will prevent this problem.

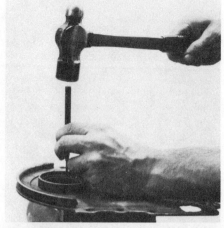 

Removing front seal from 351C front cover. Drive seal out front of cover by working around seal backside with a punch and hammer. If yours is a 429/460, drive seal out the back in similar manner, however be considerably more careful because the cover is cast aluminum.

Support front cover on a firm flat surface and drive seal into place with steel, aluminum or brass tubing which matches the seal's outer shell. Drive seal into a steel cover from the front side and into an aluminum cover from the back. Be sure the seal lip points rearward as installed in the engine.

metal cover. You can tap the seal in by working around its shell gradually, however this method is somewhat chancy. You can also use a vise with the aluminum cover, however a cylinder is required between the vise jaws and the seal so the vise will clear the cover. Gradually force the seal into place by tightening the vise jaws. Regardless of the type cover you have, *the seal lip must point toward the engine when it is installed.*

Prepare the Damper or Spacer Sleeve—If you haven't cleaned your damper yet, do it now. The main area to inspect is the rubber bond between the outer ring and the inner. It must be secure. Try to force them apart longitudinally, or back and forth. If there is noticeable movement, the bond has failed and the damper should be replaced. Also inspect the inner bore and front-cover sealing surface. Give the sealing surface of the 351C and 351M/400 damper nose the same treatment you gave the crankshaft bearing journals, page 61, but don't polish it. Some *tooth* is required to carry oil for lubricating the seal.

The 429/460 uses a separate spacer between its damper and the crankshaft sprocket. 351C and 351M/400 spacers are an integral part of their dampers. The separate 429/460 spacer provides the surface which rides against the front cover seal on these engines. Give the spacer the same treatment you gave the crankshaft bearing journals, page 61.

If your damper nose or spacer seal surface is badly grooved or unfit for sealing for one reason or another, you'll have to replace the damper spacer, or better yet, install a *sleeve* over the original seal surface. This relatively inexpensive part is available at most automotive parts stores. The sleeve is approximately 0.030-inch thick. The lip-type seal will comply with the additional thickness, so you can use a stock seal when using a sleeve.

To install a repair sleeve, clean the original seal surface and remove any burrs. Coat it with silicone sealer to prevent oil leaks between the sleeve and the damper or spacer. If you have a 351C, 351M/400, support the damper so the rubber bond will not be loaded and drive or press the sleeve on. Place a flat plate—steel, aluminum or wood—across the end of the sleeve as it sits on the end of the damper or spacer, to spread the load. Make sure the sleeve is started straight and install it until it's flush with the end of the damper nose or spacer. Wipe the excess sealer off and your seal surface is as good as new.

Water-Pump-to-Cover—Before installing

If your damper or damper-spacer seal surface doesn't clean up after giving it the same treatment you gave your crankshaft journals, or it's grooved like this one, you don't have to lay out money for a new damper or spacer. Yours can be repaired by installing an inexpensive repair sleeve.

Apply silicone sealer to the sleeve's ID, then drive it over the original seal surface. Using a wood block between the repair sleeve and hammer. Wipe away the excess sealer and your damper or sleeve will be like new.

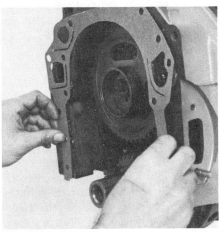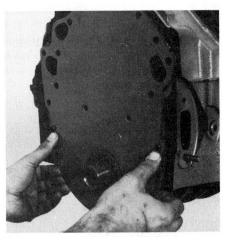
351C and 351M/400 front-cover installation. After applying sealer, install front-cover gasket. Gasket and cover locate over two dowel pins, automatically centering cover on crankshaft. Dowels can be seen directly above my thumbs as I'm installing cover. Don't forget sealer between cover and gasket. Oil and coolant are being sealed.

the front engine cover, you should install the water pump to the cover first, particularly if you have the 351C, 351M/400. The bottom cover bolts won't pull the upper section of the sheet-metal cover against the block firm enough for a good seal. Consequently, if you don't have the pump in place when installing the front cover, you'll risk oil and water leaks. Remember, the majority of the water-pump bolts thread into the block for securing the pump *and* the upper section of the cover. One exception is when using slow-curing sealer such as RTV silicone, then install the water pump immediately after you have the front cover on.

Aluminum and sheet-metal front covers have threaded holes for mounting water pumps. You'll have one gasket for the 351C, 351M/400 water pump. Apply sealer to both gasket surfaces, then install the water pump. 429/460 water pumps

Centering 429/460 front cover before tightening its bolts. After oiling damper-spacer seal surface, slide spacer into front-cover seal. This centers cover so you can secure its bolts.

ENGINE ASSEMBLY 109

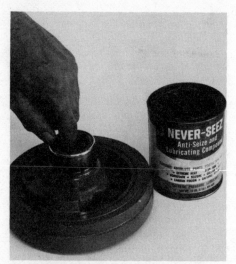  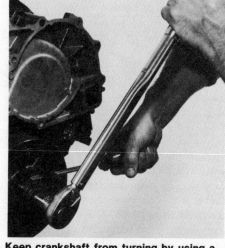

To ease future repairs, apply some anti-seize compound to the damper ID. Oil its seal surface and you're ready to install it.

Torque crankshaft damper bolt 70—90 ft.lbs. Keep crankshaft from turning by using a screwdriver between the block and one of the crank's counterweights. If the oil pan is in place, a bolt in the damper and a screwdriver can be used to hold the crank from turning.

use two gaskets, one between the pump's cover plate and the pump, and one between the water-pump assembly and the front engine cover. After you've installed the pump's cover with its gasket and sealer, install the assembly to the front cover, again with the gasket and some sealer.

Install the Front Cover—Lubricate the fuel-pump cam with some moly grease and wipe off any excess oil from the block's front-cover mounting surface. Use thinner to get it really clean so the sealer will seal.

Two dowel pins in the 351C, 351M/400 block locate the front cover whereas the 429/460 cover must be centered on the crankshaft using the damper spacer. If it isn't, the damper will probably end up off-center resulting in uneven seal loading and an oil leak.

Center the aluminum cover by installing the front cover loosely. Using the spacer installed on the crank nose and inserted in the cover seal for centering, secure the cover to the engine. Torque front-cover bolts 12—18 ft. lbs. for both styles of front covers.

A tip is in order here if you have a 429/460. Now's the time to install the thermostat-bypass hose. Regardless of age, I always replace it because it's so much easier to do now than a month or two years later. It's also easier to install now than after you have the water pump/front cover in place. You'll need a 5/8-inch hose approximately 2-3/4-inches long. This is a bit longer than required, so hold your water pump and cover in place, then measure exactly how much length you'll need. Measure from the front of the intake manifold to the back of the water pump.

After trimming the hose to length, install it, *with its two hose clamps,* on the manifold. You can now fit it to the water pump as you install it and the front cover.

Install the Damper—Apply anti-seize compound to the damper bore. It acts as a lubricant during the installation so the crankshaft and damper won't gall. It also helps in the event the damper has to be removed at a later date.

Lubricate the front-cover seal. Install the damper after making sure its key is in place. You'll be installing the damper's bolt and washer for the last time. Use it to pull the damper on, and torque it 70—90 ft.lbs. You'll find that you can't begin to apply this much torque without turning the crankshaft, so use a large screwdriver to keep it from rotating. Put the screwdriver between the bottom of the block and the front of the counterweight.

OIL PUMP & PAN

Now's the time to seal the bottom of your engine. If it's not already upside down, roll your engine over on its back. Before installing the pan, you'll have to install the oil pump and its driveshaft—it's just a tad difficult afterwards.

Oil-Pump Pickup—If you have a new oil pump or if you removed the pickup from your old one, you'll have to install the pickup before putting the pump on the engine. Pickups are not included with new or rebuilt pumps.

The 429/460 pickup tube press-fits into the bottom of its pump housing. To remove the pickup, lightly clamp the pump in a vise. I clamp it at the mounting flange. Simply rotate the pickup and gradually work it out of the pump body. A 351C or 351M/400 pickup can be unscrewed from its pump. Just remember its position so it can be installed correctly.

Rather than using a common punch, a better tool can be fabricated from a short piece of 3/4-inch pipe. Using your hacksaw, remove a 2-inch section from one end of the pipe like the one pictured. You'll now have a tool which can bear on 180° of the pickup-tube bead. Consequently, the chance of pickup-tube damage is considerably reduced. Position the pickup at a 45° angle to the pump so it will be centered in the oil pan. You can make final adjustments after installing the pump assembly.

A 351C or 351M/400 pickup should be installed so it will be level when in place on the engine, or parallel to the mounting flange. Without using sealer, thread the pickup into the pump housing and bring it around into position as it tightens. Again, make final adjustments once the pump assembly is installed.

Install the Driveshaft—I always replace the oil-pump driveshaft because of its critical function. However, if you plan to reuse your old driveshaft, look it over. If there is any appreciable wear other than the shiny spots on the drive-sides of the hexes; replace it.

With the little Tinnerman collar—the stamped-metal ring—pushed down over the shaft and your engine on its back, install your oil-pump driveshaft. Insert the end of the shaft with the collar in the oil-pump-shaft bore. This collar performs a very important function. It prevents the shaft from being pulled out of the oil pump when the distributor is removed. The importance of this feature can be fully realized by those unlucky individuals who removed the distributor only to hear an ominous clunk in their oil pan—the oil-

Install water pump before the sealer dries. If you're building a 429/460, make sure you install the by-pass hose while installing the pump. It's nearly impossible after the pump is in place.

Don't forget timing pointer. 351C and 351M/400 pointer locates over right-hand front-cover pin and is secured by two bottom cover bolts on that side. Upper end of 429/460 pointer is secured to cover with a self-tapping screw and bottom with a cover bolt.

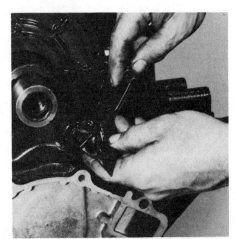

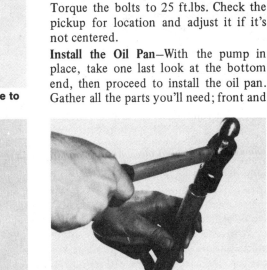

429/460 water pump, its gaskets and backing plate. Seal and assemble backing plate to pump with its gasket, then install pump on front cover with remaining gasket.

pump driveshaft. Even worse than hearing the clunk is not discovering that the driveshaft is no longer in place after installing the distributor. The amount of engine damage is proportional to the time it takes to discover there is no oil pressure.

Install the Oil Pump—You'll find your pump-to-block gasket in the engine gasket set. *Don't use sealer* when installing the pump, only the gasket. With the gasket in place on the block, install the pump while engaging the driveshaft with the pump and secure the pump to the block with its two 3/8-16 bolts and their lockwashers. Torque the bolts to 25 ft.lbs. Check the pickup for location and adjust it if it's not centered.

Install the Oil Pan—With the pump in place, take one last look at the bottom end, then proceed to install the oil pan. Gather all the parts you'll need; front and

Removing pickup from this soon-to-be replaced 429 oil pump. Note how pickup is positioned in pump, then gradually work pickup out by rotating and lifting up on it. Simply unscrew a 351C or 351M/400 pickup from pump.

Short section of 3/4-inch pipe cut like this makes an excellent pickup-installation tool.

Pickup being installed in its new pump using the "installation tool." 429/460 pump is resting on vise in which it is lightly clamped. Position pickup as it was originally, then drive it into place by working around the bead.

ENGINE ASSEMBLY 111

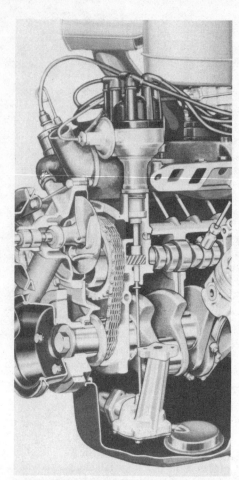

Front section of a small-block Ford shows the same oil-pump driveshaft setup as the 335 and 385 series engines. Collar at *distributor-end* of driveshaft prevents shaft from being pulled from pump and falling into oil pan when distributor is removed. Drawing courtesy Ford.

With gasket in place, install oil pump and driveshaft. No sealer is needed on this gasket.

429/460 oil-pump pickup centered as viewed from underneath. Position 351C or 351M/400 pickup so it will be level with oil-pan bottom. Make final pickup adjustments after installing oil pump.

Using a large adjustable wrench to position 351C oil-pump pickup.

With end seals located on front cover and rear main-bearing cap, fit cork pan gaskets. Projections from cork-gasket ends fit between end seals and block at front and rear cover. A small dab of sealer at the four joints readies the engine for the pan.

rear pan seals, right and left cork gaskets and the attaching bolts. You'll need 19 1/4-20 x 5/8-inch-long bolts for a 429/460, otherwise you'll need 16. You'll also need 4 5/16-18 x 3/4-inch-long bolts with lockwashers, to go directly adjacent to the front and rear pan seals. The 4 larger bolts pull the pan down against the resisting force of the front-cover and rear-cap oil-pan seals. The force required to do this would strip a 1/4-20 thread.

With the block upside down, apply adhesive-type sealer to the block's gasket surface, then lay the cork gaskets on the block—after you've figured out which one goes where and in which direction. After the gaskets are in place, put the front-cover and rear-main-cap seals into place. These rascals can be tough! First, put a little dab of silicone sealer at the end of the cork seals where the neoprene seals will join them. With the seals in place, work them down into their grooves with your thumb, particularly where they join the cork gaskets.

  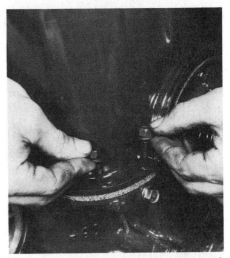

Carefully fit oil pan on block and thread all pan bolts in before tightening them. Snug them down several times as the cork gasket relaxes. Torque 1/4-20 bolts 7—9 ft.lbs. The larger 5/16-18 bolts adjacent to the front- and rear-main bearing caps: 9—11 ft.lbs.

Larger pan bolts are used at front- and rear-main bearings to provide additional force for compressing end seals.

Before setting the pan on the engine, take this opportunity to soak the timing chain with oil to give it some initial lubrication. Also, put another little dab of silicone sealer at each of the four pan-gasket and seal joints. Now install the oil pan, being careful not to move the gaskets or seals. With the pan in place, loosely install the pan bolts and *tighten them progressively*. This is a good application for a speed handle. Don't overtighten the bolts or you'll distort the oil-pan flanges. You'll notice that in going around tightening the bolts that the cork gasket relaxes after a short while. Go around tightening bolts until they begin to feel firm, then they are ready to be torqued. Check the front and rear seals to make sure they are seated in their grooves, then torque the pan bolts. Torque the 1/4-20 to 7—9 ft.lbs. and the 5/16-18 bolts to 9—11 ft.lbs. Because of the way the cork gasket relaxes, go around the pan until the bolts won't turn anymore at their specified torque.

Now that the crankcase is sealed by the oil pan, close the distributor and fuel-pump holes so a bolt, nut, wrench or any foreign object won't inadvertently end up in the oil pan. A rag stuffed in the holes does this very effectively—or some duct tape. Also, don't forget the oil-pan drain plug. Install it after you've checked its seal to make sure it's in good shape. It's always embarassing to be filling your newly rebuilt engine's crankcase with clean oil just to have it pour out the bottom of your engine at the same time.

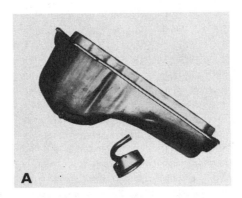 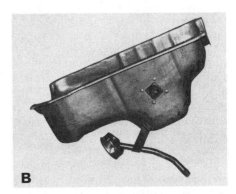
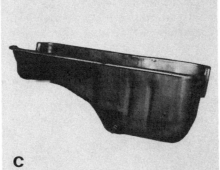 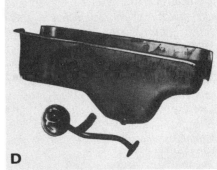

If you're swapping engines, make sure you have the oil pan and pickup for your vehicle. Here are examples of the wide differences in 335 and 385 Series oil pans and pickups: A. Passenger-car 351M/400. B. Bronco and 4x4-pickup 351M/400. C. Passenger-car 429/460. D. Econoline 460.

ENGINE ASSEMBLY 113

CYLINDER HEADS

It's now time to begin buttoning up the top side, so position your engine upright. If you are working on a bench or the floor, your engine will be unstable, so block it up under the oil-pan flanges or bell-housing bosses so the engine will sit level and be stable. If you don't do this you can count on your engine rolling over, particularly when it's unbalanced with one cylinder head installed. If you're using an engine stand, just make sure the post is tight.

Get your cylinder heads out of storage, the 2 head gaskets from your gasket set and the head bolts—20 grade-8 bolts to be exact. All 351C and 351M/400 head bolts are the same saize: 1/2-13 x 4-5/16 inches long. The top 429/460 head bolts are shorter than those for the bottom row. Their sizes are 9/16-12 x 4-3/8 inches long and 9/16-12 x 5-1/8 inches long, respectively.

Clean the Head-Gasket Surface—To ensure the best possible head-gasket seal, clean the cylinder-head and cylinder-block gasket surfaces with solvent. Check to make sure all cylinder-head locating dowel pins are in place. They are a must as they accurately locate the heads on the block deck surface. If they are missing, replace them. Do this by inserting the straight end in the block with the chamfered, or tapered end projecting outward and tap it into place. You'll feel the pin bottom in its hole when it's fully installed.

For pre-'72 engines having high compression ratios, you'll need the best sealing head gaskets available. Use Felpro's "Permatorque" or Victor's "Victorcor" head gaskets. If you use either of these gaskets, your block and cylinder-head gasket surfaces must be in the best of condition—free from distortions and smooth. These gaskets are very hard, and consequently can't conform to surface irregularities. Because head gaskets are included in a complete engine-rebuild gasket set, the set you purchase should include these head gaskets.

Coat the Head Gaskets—Head gaskets can be installed as is, however I like a little added insurance. One of the most common and easiest to appy "sealers" is aluminum paint from a spray can. Evenly coat both sides of your head gaskets with it. After the paint has dried, position the gaskets on the engine block by locating them on the dowel pins with the *FRONT designation on the gaskets to the front.* If you get the gaskets on wrong, you'll be wondering why your cylinder heads are melting and your idiot light/gage is not

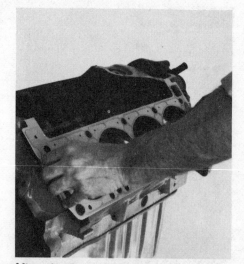

After cleaning cylinder-head and block head-gasket surfaces with solvent, fit head gaskets to block by positioning them over cylinder-head dowel pins. FRONT of gasket is indicated.

> **INSTALL TWO HEAD GASKETS**
>
> If your engine is one of the pre-1972 high-compression models, it probably has experienced detonation problems. Their small combustion chambers and resultant high-compression ratios make detonation a problem. Detonation is a real engine killer. If your engine is in this category I recommend installing two head gaskets under each head to reduce the problem. The increased clearance volume will result in approximately a one-point-lower compression ratio. So, if your engine had a 11:1 compression ratio, it will end up with 10.0:1 compression—an engine that can be operated with less tendency to detonate.
>
> Be aware that if you install two head gaskets, the rocker arms will be raised on the block with the heads approximately 0.40 inch. This is no problem unless your valves are of the non-adjustable variety. If they are non-adjustable, you may have to install longer pushrods, depending on what other work and how much of it you've done to your head. So watch out for this one.

indicating trouble. The problem is the water passages in the heads will be blocked off so coolant can't circulate through them.

Install the Heads—Flex your muscles and lift a cylinder head into position on the block. Locate it on the dowel pins, making sure the head is down against its head gasket. You don't want a head falling on the floor, especially with your feet down there. One trick to prevent this from

With fingers in an intake and exhaust port, cylinder head is carefully fitted to block over its dowel pins. Get a couple of head bolts threaded in immediately to ensure the head won't fall off.

happening is to slip a couple of the bottom head bolts in each head before installing it. Hold the bolts out of the head by stretching a rubber band between them. Then you can run the bolts in without leaving the head unsupported as you reach for a bolt. Put the other head on the block.

Oil the bolts under their heads and their threads before installing them. Torque the head bolts in sequence *and* in stages or steps as indicated in the sketch and table. Generally, head bolts need not be retorqued after the engine has been run, but it won't hurt anything to check them. 351C Boss and HO head bolts must be retorqued when the engine is hot.

Yes, 429/460 head bolts are different length. Long ones are for the bottom row and the short ones for the top row.

Oil head-bolt threads. Make sure they are clean first. Refer to the table for torque values and tightening sequence. I recommend retorquing head bolts after you've run the engine.

To keep the cylinders free of moisture and dirt, install an old set of spark plugs in your heads, some you won't mind getting broken. Secure the dip-stick tube to the front of the right head if you have a 351C or 351M/400.

VALVE TRAIN

To complete your engine's valve-train assembly you'll need 16 valve lifters, pushrods, rocker arms, fulcrums and nuts or bolts. You'll also need your squirt can full of oil or oil additive.

Install the Valve Lifters—Begin putting the final touch on your valve train by installing the valve lifters. If you have new ones, you can install them in any order; used ones must be installed in the bores they were removed from. Prime your hydraulic lifters by filling them with oil. Force oil in the side of each lifter with your squirt can until it comes out the top oil hole. After priming the lifter, oil its OD, coat its foot with moly, then install the lifter in its bore.

Pushrods Are Next—With the lifters in place, slide each pushrod into a head and on the center of a lifter. If you are installing new pushrods, they can go in either way. Install old pushrods end-for-end—the end that operated at the lifter should now be at the rocker arm. You can tell which end was which by looking at the wear pattern on the ends. The wear pattern on the lifter end will be much smaller than that at the rocker-arm end. If the ball at the rocker-arm end has worn so it is egg-shaped, replace the pushrod.

Install the Rocker Arms—Your engine will be equipped with cast-iron rocker arms, or stamped-steel ones—non-adjustable un-

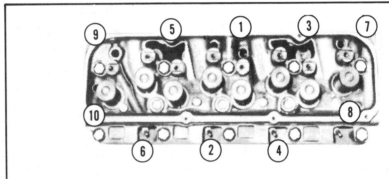

HEAD BOLT TORQUE
(ft.lbs.)

Step	351C-2V & 4V	351C Boss & HO	429/460
1	65	65	75
2	75	80	105
3	95–105 final	100	130–140 final
4		120	
5		125(hot) final	

Cylinder-head-bolt torque sequence and torque values in foot-pounds. Make sure you torque your head bolts in the steps and sequence specified, otherwise don't complain when you are faced with a blown head gasket.

less it is a 351C Boss, HO, a 429SCJ or an early 429CJ. Regardless of the type of rocker arms your engine has, relubricate the rockers prior to installation. Oil their fulcrum- and pushrod-contact points, and apply moly grease to the valve-stem tips. Begin the installation by fitting the rocker arms, pivots and nuts or bolts to the head. If you are installing your old rockers and pivots, remember to keep them together. Set the rocker arm on its pushrod and valve tip.

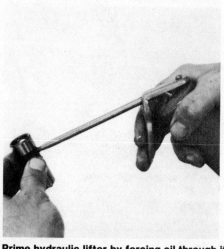  

Prime hydraulic lifter by forcing oil through its side until it comes out the top hole. Apply moly to its foot, then install it in one of the lifter bores—in its original bore if you are using the old lifters and cam.

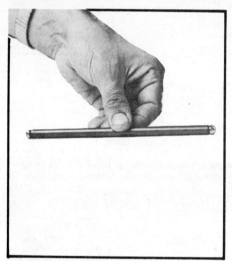

Regardless of whether your pushrods are new or not, check them for straightness by rolling them on a flat surface. I'm using a kitchen countertop. If more than an 0.008-inch feeler gage fits between the surface and the center of the pushrod, replace the pushrod. Don't try to straighten it.

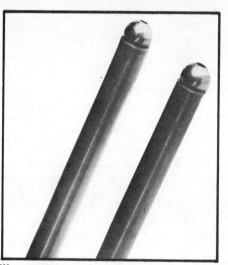

Wear pattern at rocker-arm end of pushrod extends farther around ball than at the lifter end. Swap ends when reinstalling used pushrods. As you can see, I've already installed the rocker arms loosely.

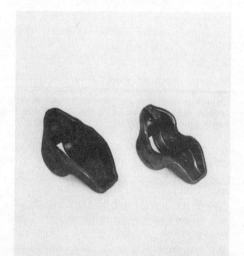

Early and late style stamped-steel rocker arms. Other than the edge trim, the significant difference between the two rocker arms is the baffle. Bent-over tab at the pushrod-end eliminates need for the separate baffle which can be installed backwards.

A valve train must be adjusted when its valve is in the fully closed position. How this adjustment is made depends on the type of rocker arms your engine has. It will have the non-adjustable stamped-steel or cast-iron rocker arms with positive-stop studs, or the rarer adjustable stamped-steel rocker arms. Non-adjustable rocker arms are adjustable, but not conveniently. Ford sells replacement pushrods which are longer or shorter by 0.060 inch for making any valve adjustments. As for the adjustable type, how the adjustment is made depends on whether you have the mechanical-lifted 429SCJ, 351C Boss or 351C HO engines—or the 429CJ with its hydraulic cam. Valves operated by mechanical lifters are adjusted so a certain amount of lash or clearance is measured between the valve tip and the rocker arm. Hydraulic lifters are designed to operate without this lash for quieter engine operation and less de-

Installing early stamped-steel rocker arms. Make sure baffle is positioned correctly. Lubricate all valve-train wear surfaces—moly on valve tips and oil on the fulcrums and pushrod ends. Install the fulcrums with their rocker arms.

Start with cylinder 1 on TDC when adjusting your valves. When adjusting hydraulic valves, first tighten fulcrum nut or bolt down to remove all slack. Rotate pushrod like this to "feel" for takeup, then rotate nut or bolt another 3/4 turn if you have the adjustable hydraulic setup. For positive-stop valves, the fulcrum nut or bolt should rotate another 3/4 to 1-3/4 turns after the pushrod *just begins* to tighten.

manding maintenance requirements. Valves seldom, if ever, require adjusting with hydraulic lifters.

Index the Crankshaft—Regardless of which type of rockers your engine has, each valve must be set when it is fully closed—its lifter must be on the base circle of its cam lobe. Both valves of a cylinder are closed when the piston is at the TDC on its firing stroke. However, at least one of the valves will be fully closed during a cylinder's other three cycles. So, rather than indexing a crank eight times to the TDC of each cylinder, the valve closings have been combined so it needs to be indexed only three times. Because the 351C and 351M/400 firing order is different from the 429/460s, the valves adjusted in the three positions will be different. Start by positioning the number-1 cylinder on TDC regardless of the engine. Refer to the chart on page 118 for crankshaft positioning and valves to be adjusted in the specific positions.

Fully Adjustable Hydraulic—Finish rocker-arm and pushrod installation and adjustment by running the adjusting nut down its stud until the slack is taken up in the rocker and pushrod. Wiggle the rocker arm to be sure all the slack is gone, but the lifter plunger has yet to collapse. When the slack is taken up, give the adjusting-nut another 3/4 turn. The valve is now adjusted and you can go to the next one. The object is to collapse the lifter plunger to the midpoint of its travel. If the lifter is fully primed, it may not collapse just yet. It will take some time before it *bleeds down*.

Fully Adjustable Mechanical—Adjusting mechanical lifters as opposed to the hydraulic type is a whole different deal. First there is *lash*, clearance measured between the rocker arm and valve tip. Lash is 0.025 inch for the 351C Boss and HO and 0.019 inch for the 429SCJ when the engine is at normal operating temperature. It's impossible to check *hot lash* now, but because the valves have to be adjusted so you can start your engine initially, you'll need to set the valves cold. *Cold lash* for the 351C and HO is 0.027 inch and 0.021 inch for the 429SCJ.

Measure rocker-to-valve-tip lash with your feeler gage. Wiggle the rocker arm to make sure there's no slack in the valve train, and adjust the nut so there's some drag on the feeler gage as you slide it between the valve tip and rocker arm. With cold lash set, loosely install the jam nuts on the adjusting nuts. You can tighten them after hot lashing.

Non-Adjustable Hydraulic—The only way to "adjust" positive-stop rocker arms or the stamped-steel rockers using the bolt-on pivots is with Ford's 0.060-inch longer or shorter pushrods. This will only be required if one of your valve seats required extensive grinding causing the valve stem to extend out of its guide excessively—this will have shown up when you were setting the installed heights of your valve springs. This situation would require a shorter pushrod if the valve stem tips were not ground to compensate. The same applies if your heads were milled, otherwise you shouldn't have any problems. Regardless of what should be, check your valves in

Setting static valve lash using a step-type feeler gage. Mechanical-cam engines such as the Boss 351C and 429SCJ will have to have their valves set now during assembly, then later when the engine can be run for hot lashing. Photo courtesy Ford.

Checking valve-tip-to-rocker-arm clearance of this non-adjustable valve train by prying between the valve retainer and rocker arm with a screwdriver to collapse the lifter.

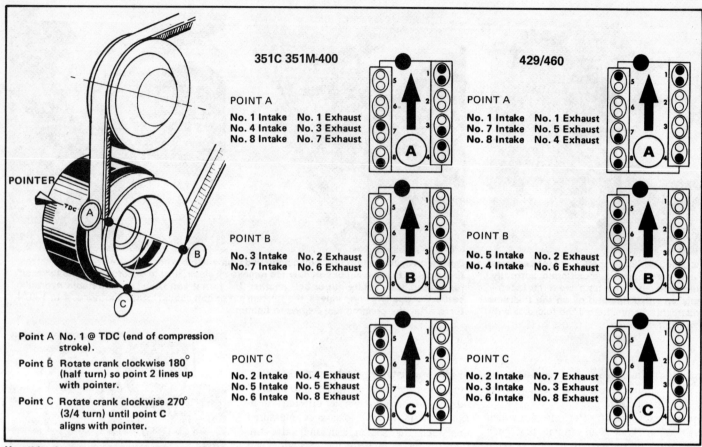

Use this chart to save time when adjusting your valves. Adjust the valve indicated in the positions shown by the black dots. If it confuses you, simply go through your engine's firing order to do the adjusting.

their closed positions.

Two methods can be used for checking non-adjustable rocker-arm adjustment. The first is by rotating the rocker arm to collapse the lifter so the rocker-arm-to-valve-tip clearance can be measured. The next, and simplest, is to count the number of turns of the nut or bolts it takes to go from *no slack* in the rocker arm to when the stamped-steel rocker-arm fulcrum contacts its stand, or when the nut contacts the shoulder on the positive-stop stud with cast rockers.

You'll need a screwdriver. Just pry between the spring retainer and rocker arm to compress the lifter *after* you've tightened down on the nut or bolt, and measure tip-to-rocker clearance. This is more difficult with the cast-iron rocker arms because the rails are not much wider than the valve tip. Consequently, you'll have to use extremely narrow feeler gages. You probably don't have them. Regardless, rocker-arm-to-valve-stem clearances should be within the following tolerance ranges: 351C and 351M/400, 0.100—0.200 inch; 429/460 using cast-iron rockers, 0.075—0.175 inch; 429/460 using stamped-steel rockers, 0.100—0.150 inch.

Just as rocker-arm-to-valve-stem clearance is directly proportional to lifter collapse, so is the vertical movement of the rocker-arm nut or bolt. The number of turns of the nut or bolt is directly proportional to how much the lifter is collapsed. This fact eliminates the need for any special tool—making the job easier and less involved.

To determine if you need different length pushrods, count the number of turns it takes to *bottom* each rocker-arm fulcrum bolt or nut. Count the number of turns from the time all the slack is taken up to when the bolt or nut *just bottoms*—not tightened, just bottomed. Wiggle the rocker arm and rotate the pushrod to check for slack, and look at the lifter to see if its plunger has moved. Run the nut or bolt down while keeping track of the number of turns. When the nut bottoms on its stud or the fulcrum bottoms on its stand, stop turning and counting. The range is 3/4 to 1-3/4 turns. If you're in the tolerance range, finish the job by torquing the nut or bolt 18—22 ft.lbs. If fewer turns are required, you need a shorter pushrod, and a longer one if more turns are needed. The object is to position the lifter-plunger approximately in the center of its travel, or 0.090-inch collapsed.

INTAKE MANIFOLD

Before covering the lifter valley with the intake manifold, pour any remaining assembly oil over the valve lifters. Pour some on the rocker arms too. You'll need three or four gaskets to seal the intake manifold to the heads and the top of the engine. All 351C, 351M/400 and later 429/469 engines use three gaskets. Two neoprene end gaskets seal the manifold to the block and one manifold baffle/manifold-to-cylinder-head gasket seals the manifold to both heads. The exception is the pre-1974 429/460 and '73 truck 460. These have the manifold heat-passage baffle directly attached to the underside of the manifold. Consequently, the manifold-to-head gaskets are separate, thus giving the appearance that there is an additional gasket. In addition to the gaskets, you'll need sealer and the attaching bolts, studs and nuts.

Put the Gaskets in Place—First step in the intake-manifold installation procedure is to fit the gaskets to the heads and the block. The cork end gaskets used to be a

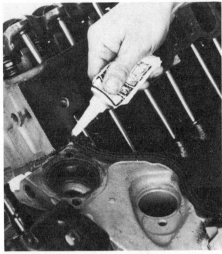

Installing the intake-manifold end seal. With a dab of silicone sealer under each end, fit the seal to the block. Another dab of silicone at each end and you're ready for the manifold-to-head gaskets.

With exception of early 429/460 engines, you'll have an integral baffle head-to-manifold gasket. Locate it on block and heads by fitting it under locating dowels at front of the left head and rear of right one. Run a bead of sealer around front and rear 429/460 water passage openings.

Don't forget sealer at the 429/460 water passages. Install the early 429/460 head-to-manifold gaskets over the manifold studs.

real pain as they were difficult to keep in place during the manifold installation, so they often leaked. This has largely been solved by making the seals from neoprene and molding *flanges* on their sides and three projections centered underneath. With the flanges hanging over the sides of the block and the projections fitted into drilled holes in the block, the seals tend to stay in place when the manifold is installed with a reasonable amount of care. To be triple-sure the end gaskets stay put and seal, apply adhesive sealer to their undersides, then fit them to the block. Apply a dab of silicone sealer to their ends at each junction of the head and the block. If you have the 429/460 with separate head-to-manifold gaskets, run a bead of sealer around the water passages at the front and rear of each cylinder head, then install the gaskets. There should be a vertical stud at the extreme end of each head. These are useful for *piloting* the manifold into place as well as positioning the manifold gaskets on the heads. Put the gaskets on the heads and fit them to the end seals. Run another bead of sealer around the water-passage openings, this time on the gaskets. Apply a dab of silicone sealer at each gasket joint. You're ready to install the manifold. On an engine using the manifold baffle/gasket, install the end seals the same way, silicone sealer at the end of the gaskets and around the water passages. Fit the manifold/gasket to the block and heads by locating it under the small pins which project from the top corners of the heads. These pins will hold the baffle/gasket in place.

Install your intake manifold by lowering it carefully onto the gaskets. Thread the bolts in and begin tightening them. Keep an eye on the gaskets to make sure they stay in place. Tighten the bolts

ENGINE ASSEMBLY 119

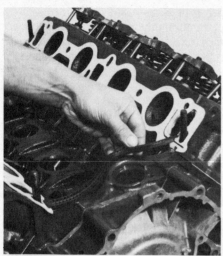

429/460 end seal going on after head-to-manifold gaskets. Silicone ends so they seal each other, then apply a small dab of sealer at four gasket and seal junctions to ready the engine for the manifold.

Carefully lower intake manifold onto engine, particularly if you have 429/460 with separate head-to-manifold gaskets. Manifold studs help guide manifold into place.

Torquing intake-manifold bolts. 351C and 351M/400 5/16 bolts are torqued 21—25 ft. lbs. and 3/8 bolts torqued 28—33 ft. lbs. Torque 429/460 intake-manifold bolts 25—30 ft. lbs.

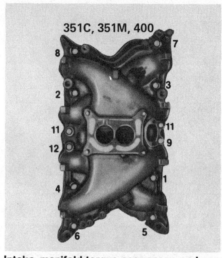
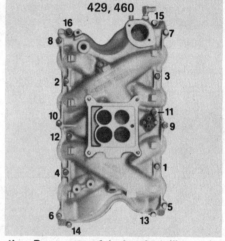

Intake-manifold torque sequences and specification. Be very careful when installing and torquing your intake manifold. Otherwise you run the risk of a vacuum leak which is very difficult to troubleshoot. Photos courtesy Ford.

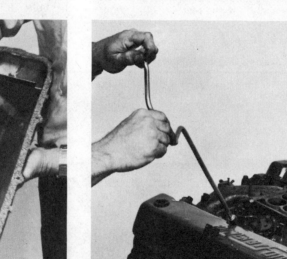

Fitting 429 valve-cover-gasket tabs to notches in valve-cover flange. As with the oil pan, valve-cover bolts have to be tightened, then torqued several times before the gasket fully compresses. Torque bolts 3—5 ft.lbs.

and final torque them in the sequence shown in the sketch. Due to its size, the 429/460 has more bolts than the 351C or 351M/400 engines.

Valve Covers—Assemble the valve-cover gaskets to their valve covers. Insert the gasket tabs in the notches of the 429/460 valve-cover flanges. Because the 351C and 351M/400 valve-cover flanges are turned up, they don't have locating notches to hold their gaskets in place. So, if you want the gaskets to stay put while installing the valve covers, use adhesive sealer between the gasket and cover. This way you can remove the valve covers without destroying the gaskets. This is particularly true if you have a 351C Boss or HO. The valve covers have to come off later for hot lashing. I recommend retorquing head bolts after running an engine, so the valve covers will have to come off for this too.

With the gaskets in place, install the covers on the cylinder heads. Secure them with their 1/4-20 bolts and lock washers. Don't overtighten the bolts as the only thing you'll succeed in doing is bending the valve-cover flanges, increasing the likelihood of an oil leak. Gradually tighten the bolts to compress the cork gasket, then torque the bolts 3—5 ft.lbs. Retorque the bolts a few times to ensure the gasket is fully relaxed, otherwise your engine will soon be leaking oil from this area.

Exhaust Manifolds—Depending on your particular engine, thin sheet-metal or composition exhaust-manifold gaskets will be in your gasket set. The sheet-metal gasket doubles as a spark-plug shield. On the right side, it is also the backhalf of the carburetor heat baffle.

A popular item at the scrap yard—right exhaust manifolds. Inspect yours closely for cracks. This one is cracked at the number 2 and 3 runners.

Your engine was not originally equipped with exhaust-manifold gaskets if your gasket set includes the composition type. Composition gaskets are a sandwich of sheet metal and asbestos. The metal side goes against the manifold and is marked MANIFOLD SIDE. I recommend not using this gasket unless your manifold or head mating surfaces are not in the best of condition—warped or pitted. However, if one of the surfaces is not in the best of condition the composition-type gasket will help seal.

To install your manifolds without gaskets, first apply graphite grease to the manifold mating surfaces, or gasket adhesive if you are using gaskets—either type. Using a couple of bolts, loose-assemble the manifolds to the heads, then your hands will be free to run in the rest of the bolts. If your engine was so equipped, install the long bolts with the studs protruding from their heads beside the center two exhaust outlets. They also secure the engine lifting plates which will soon come in handy, and the hot-air shroud for the carburetor. It mounts on top of the nuts which secure the right-side lifting plate. Tighten and torque the manifold bolts from the *center outward*. Torque 351C and 351M/400 bolts 18—24 ft.lbs. and 429/460 bolts 28—33 ft. lbs. Install the lifting plates and carburetor hot-air shroud now.

Install the Thermostat and Its Housing—This is a good opportunity, and your last one, to make sure the *water-pump by-pass orifice plate* is in place in your 351C before installing your thermostat and its housing. The orifice plate *meters* how much coolant is recirculated through the engine rather than going to the radiator. If it isn't in place, the recirculating coolant will not be restricted and less coolant will

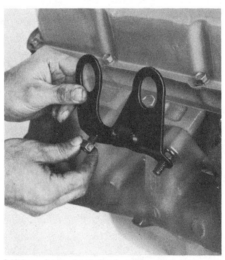
Lifting lugs installed with the exhaust manifold. I prefer not using a gasket between the manifold and head, but if you insist on installing those in your gasket set, put the metal side against the manifold, not the head.

Installing 351C heater-hose inlet tube. Old one was badly rusted so I removed it by wrenching it out with Vise-Grip pliers. Do the same if yours is bad.

pass through the radiator to be cooled. The result: an overheating engine. So make sure your 351C has this restriction. On the other engines, the restrictor is built in, so you don't have to worry about it.

When installing your thermostat in the block, or in the intake manifold in the case of 429/460 engines—they have *wet manifolds*—make sure you have it pointed in the right direction. One end should indicate TOWARD ENGINE. Install this end *down* in the engine block or intake manifold. Now you're ready for the housing. Check it for excessive rust if it's the steel type used on 351C and 351M/400 engines. These parts literally *go away* if water is used for coolant. Replace it if it's pitted badly. To finish installing the thermostat and housing, seal the gasket to the block/manifold. Apply sealer to the housing and install it on the block/manifold. Torque the bolts 12—15 ft.lbs.

MISCELLANEOUS HARDWARE

Vacuum Fittings—Intake manifolds are equipped with one or more vacuum fittings to supply various accessories and emission-related devices with a vacuum source. Due to their wide variances and the resulting complexities involved in covering them, you'll just have to remember how and where yours are installed and used.

Sending Units—Two engine sending units are used: oil pressure and water temperature. The oil-pressure sender threads into the engine block behind the intake manifold. Seal its threads and run it firmly

Install thermostat in the right direction. It locates in a counterbore. Apply sealer to the housing and gasket, then install the housing.

If your heater-hose fitting looks like this, replace it. It's another example of using water in a cooling system instead of antifreeze.

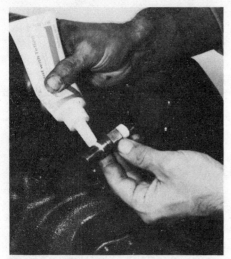

Preparing 429 water-temperature sending unit for installation. It goes in the intake manifold. 351C and 351M/400 senders locate in the front of the block.

Installing oil-pressure sender after sealing threads. If your engine was oily at the backside prior to rebuilding it, suspect the oil-pressure sender and replace it.

into the block. If it's the idiot-light-type sender, replace it if the top-rear of your engine was wet with oil. The oil sender was probably the cause as they are notorious leakers.

Water-temperature sending units install immediately below the thermostat in 351C and 351M/400 engines, and in the intake manifold of 429/460s. Install yours with sealer on the threads.

Fuel Pump—Put the fuel-pump gasket on the pump with sealer. Run a bead of sealer around the engine side of the gasket too. Prelube the pump actuating arm with moly, then install it. How you install the pump depends on whether your engine is a 351C, 351M/400, or a 429/460. The 351C and 351M/400 has a *vertical* bolt pattern with a bolt at the top and a stud and nut at the bottom. This is handy because you can easily position the pump over the stud, then hold it in place with the nut against the force of the pump arm while you install the bolt. Finish the installation by torquing the nut and bolt

Lubricate fuel-pump-lever wear surface with moly, then install it with its gasket. One bolt and one stud with a nut retain 351C and 351M/400 pump whereas two bolts are used with 429/460.

12—15 ft. lbs. 429/460 has a horizontal bolt pattern. With its gasket in place on the pump, moly on the arm and one bolt in hand and the other in reach, locate the pump in position and secure it by threading the bolts in. Tighten the bolts, then torque them 20—24 ft. lbs.

Carburetor—Carburetors have varied widely since 1968, from the performance-oriented 351C Boss and HO and the 429CJs and SCJs to the current emissions/mileage tuned carburetors. Also, some do and some don't use carburetor spacers. There are so many combinations that being specific would be too complex, however they are generally the same except for hookups for the various emissions-related devices, both on and off the carburetor. There are also considerable differences in carburetor spacers, from those which insulate the carburetor base to those that recirculate exhaust gas back through the induction system. Spacers vary from a thin stainless-steel plate to an aluminum spacer with an EGR (exhaust gas recirculating) valve attached. The order of installation from the carburetor up is gasket, spacer, gasket, then the carburetor.

Carburetor spacers don't present any particular problem except for one, and it can be a big problem. The one I'm referring to is the aluminum spacer/EGR valve. The aluminum eventually burns out from the hot exhaust gases, resulting in exhaust noises coming from the carburetor! Check yours closely for signs of burning out, then replace it if it appears to have started. As a result of the burnout problem, this spacer was changed to cast iron as have all the replacement spacers. A new one from your Ford dealer should be cast iron.

With the gaskets, spacer and studs on the intake manifold, add the carburetor. Before installing the 4 hold-down nuts, connect the choke heat tube. Thread the nuts on their studs and tighten each one *a little at a time* until all are at the correct torque—don't exceed 15 ft.lbs. The nuts are easily stripped and progressively tightening them eliminates the possiblity of breaking a flange off the carburetor.

As you can tell by the number of parts laying around, you can install a few more before your engine is in the engine compartment—some will have to be. Let's get on with the actual installation. Most of the remaining parts and components are easier to install just prior to the time your engine is ready to be dropped into place—while the engine is hanging in mid-air.

Ford does a landmark business with these aluminum EGR spacer plates. They burn out with great regularity. This one is also plugged up with carbon, so it didn't work anyway. Inspect yours for signs of burning and replace it if it looks bad.

Install choke heater before carburetor. 351C is unique in that two center, right manifold bolts also retain outboard side of heater plate. This was changed because many attempts were made at removing manifolds with these bolts in place as you can see on page 41.

With gasket or spacer and gaskets in place, set your carburetor onto intake manifold. Attach choke heat tube to choke *before* securing carburetor, otherwise you'll have a problem connecting the heat tube.

Tighten carburetor nuts with care. Don't exceed 15 ft.lbs.

8 Distributor and Carburetor Rebuild

Due to the general lack of information on distributor rebuilding, I'll show how to determine if yours needs rebuilding, then show you how to do it. I'll also touch on rebuilding the Autolite/Motorcraft 2V carburetor.

DISTRIBUTOR TYPES

Two basic types of distributors are installed in Ford engines: *breaker-point* and *breakerless* (solid-state). The difference is in how they produce a spark. The breaker-point distributor opens and closes a set of breaker or contact points on a breaker plate. The breakerless distributor accomplished the same thing electronically through a magnetic-triggering device and an amplifying module.

Otherwise, the two are basically the same. Both are driven at 1/2 crankshaft speed by a camshaft gear through a shaft which runs in bushing/s in the distributor housing. Distributors installed in engines prior to 1970 may or may not use two replaceable bushings. Beginning in 1970, all distributors use the top bushing only and depend on the end of the distributor shaft running in the cylinder block to support the gear end.

At the upper end of the distributor shaft a cam operates breaker points, or an armature for triggering the magnetic pickup. The cam or amature is free to rotate on the shaft within limits controlled by centrifugal-advance weights which change spark timing with RPM. Above the cam or armature is a rotor. The same kind is used on both type of distributors. As it rotates, it completes the circuit between the coil and each spark plug in the correct firing order. The distributor baseplate mounts the breaker plate or the magnetic-pickup assembly. Otherwise the two distributor types are the same.

All distributors installed on all 351C, 351M/400 and 429/460 engines have *dual-advance*. Spark timing is varied by centrifugal-advance weights that rotate the cam or armature and one or two vacuum diaphragm/s rotate the breaker plate or magnetic pickup assembly around the center shaft. One exception is the

Solid-state distributor with centrifugal- and vacuum-advance components. Other than the solid-state's armature, stator and pickup-coil assembly versus the cam and breaker-plate assembly of the point-type distributor, the distributor assemblies are basically the same. Drawing courtesy Ford.

Checking a distributor the right way—on a distributor test stand. Centrifugal advance is being adjusted by bending the advance-stop tab.

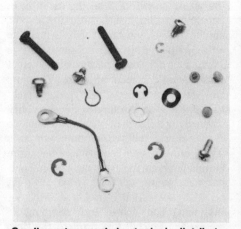

Small parts used in typical distributor assembly. Be careful when disassembling yours so you don't lose any. Hours can be spent finding replacements.

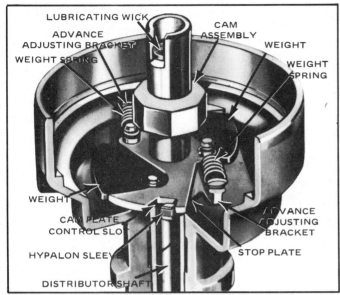

Centrifugal weights, springs and slots establish rate and amount of spark advance. Number stamped adjacent to each of two advance-limiting slots is maximum centrifugal advance when that slot is located over advance-limiting tab—10L is 10°, 13L is 13°, etc.

Not all distributors are equipped with this, a radio-interference supression plate. It lifts off after cap and rotor are removed.

dual-point distributor used on some 351C-4Vs with 4-speed transmissions and all 351C Boss, HO, 429CJ and SCJ engines.

Another variation occurs in the type of vacuum diaphragm. All pre-1968 distributors use a *single diaphragm* operated by carburetor-venturi vacuum. Beginning in 1968, the *dual diaphragm* was used for emission-control reasons on engines sold in some areas. This unit uses the carburetor-vacuum actuated diaphragm. It also has a *vacuum-retard* diaphragm in tandem with the advance diaphragm to retard the distributor timing when there is high manifold vacuum—at idle and during deceleration when the throttle is closed. Other than the diaphragm, the two distributors are mechanically the same even though they have different advance/retard curves.

All solid-state distributors use the dual diaphragm. 1977 and later models have a larger diameter distributor cap with an adapter between the distributor housing and the cap. These also use a unique rotor and wiring-harness leads with spark-plug-type connectors at both ends.

Does It Need Rebuilding?—First determine whether your distributor *needs* rebuilding before you tear into it. Remove the cap and rotor, then wiggle the upper end of the shaft back and forth—or at least try to. If it moves a noticeable amount, something is worn too much. It could be the ID of the cam or sleeve, or the upper shaft bushing. Of course, it's possible they are both worn. To isolate the two, remove the cam from the end of the shaft using the procedure on page 127, then try wiggling the shaft. If it still moves, you know it's the bushing. If it doesn't, then it's the cam or sleeve ID. The problem could be a worn shaft, but it'll probably be the bushing or the cam or sleeve. To make sure it's not the shaft, mike its upper portion that runs in the bushing and compare this figure to what you get when miking an unworn section of the shaft. To do this you'll have to remove the shaft, so refer to the teardown section later in this chapter. First I'll describe a "real-world" way of determining if your distributor needs rebuilding.

Let Performance Be the Judge—Let your distributor's performance be the determining factor as to whether it needs attention or not. To do this, take it to a shop which has a distributor test set and knows how to use it. Check for correct centrifugal advance and vacuum advance/retard—applies to both types of distributors—and dwell variations for the breaker-point type. Chances are your centrifugal-advance mechanism will check out OK. If it doesn't, a minor adjustment should correct the problem. Possibly the weight pivots need to be cleaned and lubricated.

The vacuum advance/retard mechanism is a more likely cause of malfunction. This usually occurs in the form of a leaky vacuum diaphragm, meaning it will have to be replaced. With the breaker-point distributor, another possibility is the movable breaker-plate may be hanging up. This will either be caused by a bent breaker-plate, or the three nylon buttons on which the plate slides are worn. You may be able to straighten the plate, but worn buttons mean breaker-plate replacement—you can't buy new buttons.

You also need to determine the condition of your distributor's upper distributor-shaft bushing and the breaker-plate pivot. This is done by monitoring the *dwell* on the test set. If dwell varies, the cam is moving in relation to the points, or the points are moving relative to the cam. When the cam and points are farthest from each other, the points stay closed longer, thus *increasing dwell*. The reverse is true when they come closer together. The unlikely cause of this is a worn upper bushing, or possibly excess wear between the cam and the distributor shaft. More than likely it's the bushing. Although it's much more durable, the solid-state distributor can exhibit the same characteristics as the breaker-point type when the top bushing is worn. The armature will move in relation to the pickup, causing erratic dwell readings.

Breaker-plate-pivot wear can also cause dwell variations. A brass bushing is pressed into the movable breaker plate. The breaker plate pivots on a pin attached to the baseplate. Instead of the cam moving in this instance, the points move in relation to the cam, but the effect is the same. So, if the breaker-plate pivot is worn so there is dwell variation, the *breaker-plate assembly* should be replaced. The bushing is not serviced separately.

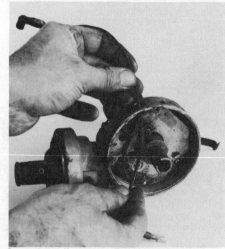

Getting the last of the familiar hardware out of the way, points and condenser.

Remove the primary wire by sliding it *in* through the distributor housing.

Keep a finger on C-clip when removing it. If one flies off you can just about forget finding it. Lift diaphragm off post, then slide diaphragm assembly off distributor housing.

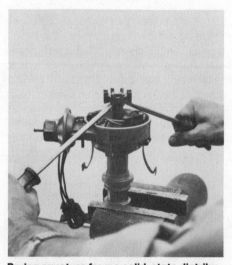

With two attaching screws removed—ground wire is retained by one—lift out the breaker-plate assembly.

Prying armature from a solid-state distributor. Two screwdrivers and light pressure should do it. Be careful not to lose armature-locating roll pin.

REBUILDING YOUR DISTRIBUTOR

Before tearing down your distributor, make sure your Ford dealer has the bushing/s in stock—and that he'll sell them to you. You can thank the Environmental Protection Agency for this if he won't. He'll prefer selling you a new or rebuilt distributor rather than the bushing/s.

Buy upper bushing B8QH-12120-A for single-point distributors and C5AZ-12120-A for dual-point distributors. If you need it, get the lower bushing C5AZ-12132-A. If you encounter problems with obtaining these, you may have to start shopping for a replacement distributor. Assuming you can come up with the required parts, let's get on with the job.

Disassembling the Breaker-Point Distributor—After removing the cap and rotor, remove the breaker points and condenser. Keep the small parts, nuts, screws, clips and the like in a can, small box or paper sack—anything to keep them from getting lost. It's literally impossible to find replacements. Remove the vacuum-diaphragm assembly after disconnecting its link from the breaker plate. Use a small screwdriver to pry the clip off and hold a finger on the clip so it doesn't fly away.

The breaker-plate assembly can come out in one piece, but I've shown it being removed in pieces. To remove the plate, remove the C-clip, flat washer and spring washer from the pivot, then lift the breaker plate out. This exposes the baseplate which you can also remove. Don't lose the three nylon buttons. They pop out easily.

Check the breaker-plate pivot for wear by putting the two plates together and try to slide the plate sideways. If you can feel movement, replace the complete assembly.

Disassembling the Breakerless Distributor—If you have a breakerless distributor, tear it down similarly. Remove the cap and rotor. On '77 and later models remove the cap adapter too.

Remove the *armature*—the wheel with eight vanes—by prying it off the shaft with two screwdrivers. Don't lose the *roll-pin* which keys the aramature to the shaft. Remove the vacuum-diaphragm assembly by unhooking it from the magnetic-pickup assembly. Use your fingernail to remove the wire retainer, then lift the pickup assembly off the baseplate. The baseplate can be removed now by undoing the single remaining screw.

With the baseplate out of the way, your distributor looks nearly identical to a breaker-point type in the same state of disassembly. The difference is the straight

Magnetic-pickup retaining clip is best removed with your fingernail. Remove vacuum-advance diaphragm and lift out pickup.

Push centrifugal-weight C-clips out of their grooves to remove weights. Remember to keep a finger on clip while you're removing it.

If centrifugal weights are retained with this type clip, don't attempt to remove them. You'll destroy them and they aren't serviced—meaning you can't buy them.

Breaker-point-cam and armature-sleeve removal is the same for breaker-point and solid-state distributors. With lubricating wick out of the way, remove retaining clip. You'll need small needle-nose pliers for this. Cam or sleeve assembly should lift off shaft easily. However, some coaxing may be required if shaft is varnished. Remove thrust washer next. This one has ears for locking it in place on shaft plate.

Support backsides of gear hub or collar to drive out roll pins. I'm using a 1/8-inch pin punch. A nail with the end ground flat will work.

Cutting a short section of the old oil-pump driveshaft to use as a punch for removing gear and collar. It's easier to use than the full length of the shaft.

DISTRIBUTOR REBUILD 127

Supporting backside of gear to drive shaft down through gear with piece of old driveshaft. Soft jaws of this vise won't damage the gear. Distributor housing will support load required to remove collar. It's a relatively loose fit.

Long skinny punch being used to remove bottom distributor-shaft bushing. Support housing base and leave room for bushing as it is driven out.

sleeve which accepts the armature rather than the cam used for operating a set of points. The remainder of the teardown is the same for both types.

Disassembly From the Baseplate Down—You can remove the main distributor shaft at this point, but I've shown removing the centrifugal-advance weights and springs first. Before you remove the springs, weights or cam or sleeve, mark them so you'll have a reference for reinstalling them in their original positions. A dab of light-colored paint on *one spring and its two attaching points* and *one of the weights and the pin it fits* will do the job. The cam or sleeve has two different-length notches at the edge of its plate to limit maximum centrifugal advance. The notch with the correct advance is located over the tab which is fitted with a Hypalon sleeve. Put a dab of paint at the edge of this slot. You have to be careful not to remove the paint while you're cleaning the parts.

Cam, Spring and Weight Removal—When unhooking the centrifugal-advance return springs, stretch each of them *only* far enough to lift them off their pins. The cam or sleeve can be removed after you have the springs off. Remove the retainer from the cam or sleeve center. Lift out the lubricating wick. Use small needle-nose pliers to lift one end of the spring out of its groove and a screwdriver to keep it from slipping back in while you lift the other end out. After removing the retainer, lift the cam or sleeve off the shaft. Don't lose the thrust washer that's located directly under it.

Now for the weights. If retained with C-clips, pop them off and remove the weights. If plastic clips are used, *don't attempt to remove them*. They look like flat washers with fluted IDs, or fingers, on their inside diameters. Ford dealers don't stock them. If you remove them they'll break and you won't be able to replace them. If this sounds like bad news to you, it shouldn't. Centrifugal-pivot wear is not a problem. When they do wear the effect is minimal. Just make sure the weights pivot freely.

Distributor-Shaft Removal—We're to the state of disassembly where all distributors are the same. Remove the gear and collar. Back up the gear hub—not the teeth—with the jaws of a vise or something that will clear the roll pin as it is being driven out. Use a 1/8-inch-diameter pin punch, or a drift punch no larger than 1/8 inch, 3/4 inch from its end. Remove the pin from the collar.

Press or Drive the Gear and Collar Off—If you don't have a press, you'll have to use some less-sophisticated means to remove it. Regardless of which method you use, back up the gear with something fairly substantial. Two parallel steel plates will do if you have a press or slightly open vise jaws will work. Suspend the distributor upside-down by supporting it squarely on the backside of the gear. Don't press or drive directly on the end of the shaft. Use a short section of your old oil-pump shaft inserted in the end of the shaft. As the shaft end nears the gear, be ready to catch the distributor housing.

Remove the Bushing/s—If your distributor has the short lower bushing, you'll need a 3-1/2-inch-long, 3/8-inch-diameter punch to knock it out the bottom of the housing. Support the housing on its base and over a hole or slot so the bushing will clear as it is being driven out.

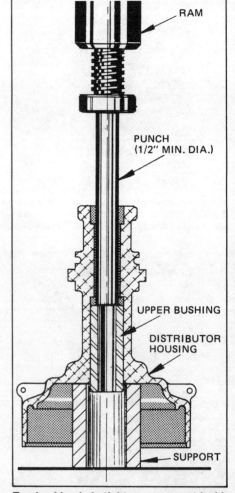

Top bushing is in tight, so a support inside housing is needed to avoid housing damage. A 3/4-inch deep socket can be used for the support. You'll need at least a 1/2-inch punch to catch the bottom lip of the bushing.

Real-life removal of top distributor-shaft bushing with a hydraulic press. Note support under housing.

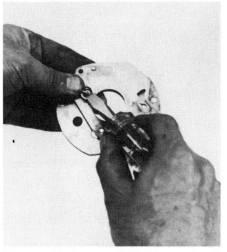

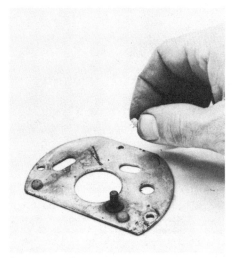

Disassemble breaker plate by removing C-clip from pivot. There'll be a flat washer and wave washer under C-clip and three nylon buttons between movable and fixed plates. Be careful, they are easy to lose.

To remove the top bushing, invert the housing and support it as in the sketch. It's advisable to remove this bushing with a press because of the additional force required, and because the housing could be damaged. Regardless of how you remove it, press or hammer, you need a punch large enough to bear on the bottom of the bushing, but no larger than a 1/2-inch diameter, and long enough to reach down in the housing. If you drive the bushing out, be careful!

Clean-Up and Inspection—Clean all the parts. An old brush and a bucket, or a can of cleaning solvent or lacquer thinner are perfect for this job. Be careful not to remove your locating marks from the cam, springs and centrifugal weights.

Inspect all the parts. The shaft should look pretty good, but polish it with some worn 400-grit emery cloth to give it a new look. Mike the worn and unworn shaft diameters and compare the results. If it's worn more than 0.001 inch, replace it. As for the cam, feel the peak of the lobes across their width. If you feel any distinguishable ridge from the breaker-point rub block, replace the cam. Otherwise, it's OK. Check the movable breaker-plate pivot and bushing if you haven't done so already.

REASSEMBLING THE PIECES

After cleaning, inspecting and obtaining any parts you'll need for reassembling your distributor it's time to retrace your steps. Basically all that has to be done is to reverse the teardown process, but with considerably more care.

Install the Shaft Bushing/s—Begin distributor assembly by installing the bushings in the housing. Install the bottom one first—Ford C5AZ-12132-A—by standing the housing upside down and driving or pressing the bushing squarely in its bore using a 5/8-inch diameter punch. When the bushing feels solid as you're driving it in, it should be at the bottom of its counterbore and flush with the bottom of the housing.

Turn the housing over to install the top bushing—Ford B8HQ-12120-A, or C5AZ-12120-A with dual points. Lubricate the bushing and center it over its bore. Use a punch between the press ram or your hammer, and drive the bushing in *straight* until it stops at the bottom of its counterbore.

Burnish the Bushings—To make the bushing IDs as smooth as possible, they should be *burnished*. This is normally done with a burnishing tool, however these are not commonly available, so use a 15/32-inch diameter ream. With the ream well oiled, turn it backwards in the bushings—opposite from the direction you turn it for reaming. This will smooth out any burrs or irregularities in the bushing bore/s without removing any material.

Install the Shaft—Oil the distributor shaft and slide it into place. It should rotate freely. Secure it by sliding the collar—flanged end first—onto the shaft. Line up the roll-pin hole with the one in the shaft. It's drilled off-center, so you may have to rotate the collar to line it up. If you are installing a new shaft, you'll have to drill an 1/8-inch roll-pin hole. To set the correct end-play, drill a new hole through the collar *and the shaft* 90° to the original hole with a 0.024-inch feeler gage inserted between the collar and the housing. This will set the end-play. Drive the roll pin into place after you've lined up the holes.

Install the drive gear next. Start the hub end on *after* lining up the roll-pin holes. This should be done first because the gear will be too difficult to turn once it's on the shaft. If you've installed a new shaft, you'll have to drill a new 1/8-inch hole. Do this after you've located the gear 4-1/32 inches below the bottom of the housing mounting flange—just above the O-ring groove—to the bottom of the gear. Install the roll pin and you're finished with the bottom end of your distributor.

Assemble the Centrifugal-Advance Mechanism—Gather centrifugal-advance pieces in preparation for reassembling them. Start by spreading a *light film* of moly grease on the weight pivots, then install the weights using your marks to determine their locations. Secure the weights with C-clips.

Lubricate the upper end of the shaft by filling the grooves with moly grease, then slide the thrust washer into place over the shaft. Lubricate the bottom of the cam or sleeve and install it so the advance-limiting tab projects up through the correct notch at the edge of the cam or sleeve plate. Use your marks as reference. Check to make sure the Hypalon sleeve is in place on the tab.

Again using your marks, install the advance-return springs between the shaft and cam or sleeve mountings. Attach them at the shaft plate first, then to the cam or shaft pin. Be careful not to stretch

DISTRIBUTOR REBUILD 129

Checking breaker-plate-pivot bushing wear. Loose-assemble movable breaker plate to baseplate. If it feels loose as you try sliding plate from side-to-side and wiggling it, bushing is worn out and breaker-plate must be replaced. You can get this and some change for a ten-dollar bill at your Ford dealer. Beats laying out the cash for a complete distributor.

Position bottom bushing squarely over its bore, then drive it in until it bottoms. I'm using a soft punch to avoid damaging bushing ID.

Lubricate OD of top bushing, then drive it into place. A press would be better, but a soft punch and hammer will work. Just be careful not to damage bushing bore.

400-grit cloth being used to remove varnish and restore shaft's original polish. Oil shaft, then install it in housing.

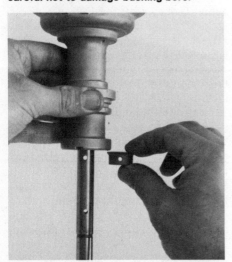

Pin holes in shaft, collar and gear are drilled off-center. Line them up before assembling them, particularly the gear. It's practically impossible to move the gear once it's on.

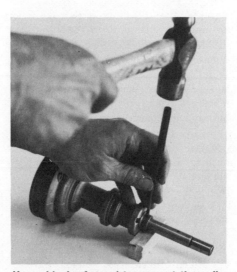

Use a block of wood to support the collar while installing its roll pin.

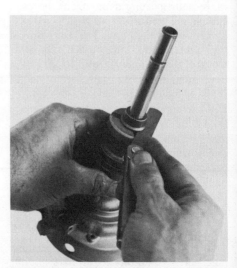

Clearance between collar and housing should not exceed 0.035 inch. I'm pushing down on the housng with the assembly supported by the shaft while clearance is measured.

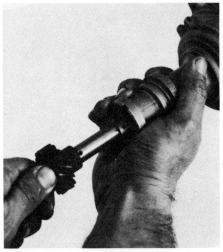
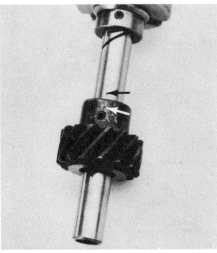

Slide gear on the shaft while keeping the roll-pin holes lined up. Use the prick-punch marks (arrows) as reference. Support the gear hub like the collar while installing its roll pin.

the springs!

Complete the assembly of the cam or sleeve and centrifugal-advance mechanism by installing the retaining clip on the end of the shaft. Feed it in between the cam or sleeve and the shaft using your pliers, then push it into place with two small screwdrivers.

Breaker-Point Distributor Assembly—Baseplate Up—Assemble the movable breaker plate to the baseplate after lubricating the pivot with moly grease and making sure the three nylon wear buttons are in place. Install the spring washer with its curved section against the top side of the bushing and the outside edge bearing against the underside of the flat washer. Secure the plate, spring washer and flat washer with the C-clip.

Install the breaker-plate assembly, movable or fixed, into the distributor housing. Rotate it until it lines up with the mounting holes in the housing, and secure it with two mounting screws. The one closest to the primary-wire hole also attaches the breaker-plate ground wire.

Finish assembling the top end of the distributor by installing the primary wire, new breaker points and condensor. Set the points by rotating the cam so the rub block is on top of one lobe. Set point gap to 0.021 inch unless you have a single-diaphragm or a dual-point distributor. Then the gaps should be set at 0.017 inch and 0.020 inch respectively. Recheck the

Install centrifugal-advance weights after lightly lubricating their pivots. Secure them with their C-clips.

Don't forget the thrust washer. If yours is this type, install the ears or tabs down in the shaft-plate holes.

Lubricate the distributor shaft and the base of the cam or sleeve assembly. Position the centrifugal weights so the cam or sleeve will seat down against the thrust washer. Make sure the correct centrifugal-advance-limiting slot is located to the tab.

DISTRIBUTOR REBUILD

Cam or sleeve-retaining clip is easier to install than to remove. Place it over the end of the shaft, then push it down into place with two small screwdrivers.

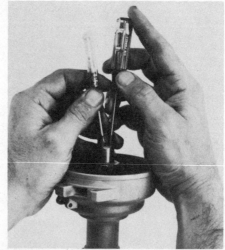

Don't forget the lubricating wick. Push it down on top of the shaft and saturate it with SAE 30 oil.

Hook each centrifugal-advance spring over its bent-up tab at the shaft plate, then shove the other end down over the anchor pin at the cam or sleeve plate.

Don't forget the Hypalon sleeve which fits over the advance-limiting tab. Push it all the way over the tab ears so it locks in place.

With the pivot pin lubricated and the three nylon buttons in place, assemble the movable breaker plate to the baseplate. This is what it should look like. Wave washer has convex side against pivot bushing with flat washer followed by C-clip on top of it.

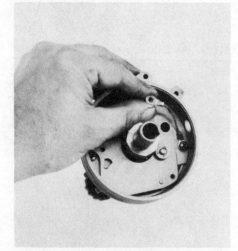
Secure breaker-plate assembly in the housing with two screws. Ground wire is installed under the screw closest to breaker-plate pivot as shown.

Vacuum-can linkage is slipped through the clearance slot at side of housing and secured to pin and housing with C-clip and two long screws.

points once they are set and secured to make sure the gap didn't change when you tightened the attaching screws. Finish the job by *very lightly lubricating* the cam with the grease that accompanied your new points, then install the rotor and cap. A light film of the grease is all that's required.

Breakerless Distributor Assembly—Baseplate Up—Install the baseplate using one screw opposite the vertical slot in the housing. Install the magnetic-pickup assembly after lubricating its pivot, and retain it with the retaining clip. Slide the wiring harness into its slot in the housing, then secure it and the baseplate to the

After installing new points and a condenser, adjust the points. Rotate cam so the rub block is on the nose of a cam lobe. Use a clean feeler gage to measure the point gap. Recheck the point setting after tightening screws. Slip the primary wire out through the distributor housing and attach it and the condenser lead to the points. Install a new rotor, cap and O-ring and you've got a new distributor.

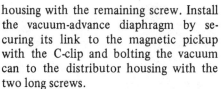

housing with the remaining screw. Install the vacuum-advance diaphragm by securing its link to the magnetic pickup with the C-clip and bolting the vacuum can to the distributor housing with the two long screws.

With the vanes pointing up, install the armature on the sleeve. Push it all the way down on the sleeve, lining up the longitudinal grooves in the armature and shaft. Install the small roll pin in the groove. Tap it into place with your hammer and punch. Install the rotor and distributor cap. The distributor is ready to be installed.

After installing a solid-state distributor base plate, lubricate the pickup pivot. Position pickup assembly on baseplate and place molded section of harness into housing slot, securing it with a baseplate screw. Secure pickup to baseplate with wire clip.

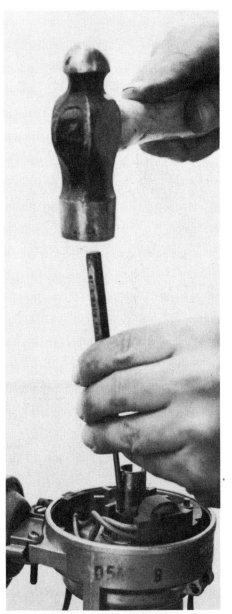

With the armature and shaft grooves lined up and armature vanes pointing up, push armature down on shaft until it bottoms. Lock it to the shaft by installing roll pin in groove. I'm tapping the pin into place with a hammer and punch.

DISTRIBUTOR REBUILD 133

Information off your carburetor ID tag is necessary when ordering a kit or parts for your carburetor.

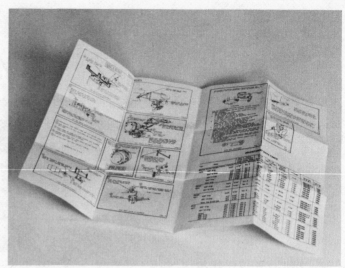
Depend on the carburetor-rebuild instruction sheet when rebuilding your carburetor.

CARBURETOR REBUILD

Carburetor rebuilding is a relatively simple job and the instructions included in rebuild kits are self-explanatory. However, before investing in a rebuild kit for your carburetor, check its throttle shaft for wear—actually the wear will be in the throttle body. The point is, if it's sloppy, a rebuild kit won't correct the problem. The cure for a worn throttle shaft is to replace the carburetor with a new or rebuilt one.

Identification—You need to know the engine, year and model when ordering a carburetor-rebuild kit. Also, there's a tag—at least there should be—under one bowl screw. This tag provides even a more explicit identification of your carburetor. Have this information with you too.

The photos and description of the rebuild are of the Autolite/Motorcraft 2100 (2-V) carburetor only. 4-V carburetors installed on many of the subject engines include the Autolite/Motorcraft 4300, Holley and Rochester carburetors. The last two were used mainly on performance engines. In-depth information on these carburetors is available in HPBooks' *Holley Carburetors & Manifolds* and *Rochester Carburetors*.

Carburetor Teardown—After you have your rebuild kit in hand, study the enclosed instruction sheet before tearing into your carburetor. I've found the blow-up drawing is extremely helpful. After having done this, start your teardown by stripping the external hardware off your carburetor; automatic choke and

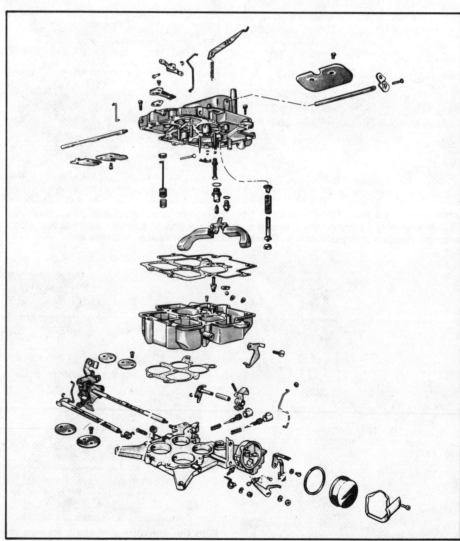
Autolite Model 4300-4V carburetor assembly. Regardless of whether yours is a 2V or 4V, the rebuild procedure is essentially the same.

Begin carburetor teardown by removing the external hardware; automatic-choke assembly, fuel filter, throttle solenoid and any other items which have been hung on your carburetor.

Don't overlook removing the air-cleaner hold-down stud before you remove the carburetor top. It threads into the main body.

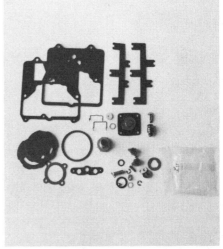
After your carburetor is torn down and cleaned, lay out the rebuild kit. Be careful. There are several small parts that are easy to lose, particularly the ball check/s. There are more parts than you need. This allows one kit to be used to rebuild more than one carburetor model.

filter—not the butterfly—and the choke vacuum can and throttle solenoid or dashpot if yours is one of the later emission-controlled models.

Remove the bowl cover. Don't forget to remove the air-cleaner hold-down bolt from the center. You now have access to the float, its needle and seat, the venturi booster assembly and main jets. Also remove the discharge-nozzle screw, accelerator-pump discharge-ball check, and nozzle-and-venturi-booster assembly. Be careful with these parts. They are easily damaged and lost.

Strip the bowl cover and throttle body of their "exterior" components. The main, or throttle body will have a power/economizer valve, idle screws and accelerator pump. A tip here: screw the idle-mixture screws *in* until they bottom, while counting the number of turns. Make a note of it for reassembly. After removing them, the cleaning process can start.

Soak the parts in carburetor cleaner or lacquer thinner—long enough to remove or soften the carbon and varnish deposits so they can be rinsed or wiped off. A note of caution: *Don't soak any rubber or plastic parts in solvent.* This includes the float. Otherwise, there's a good chance they will be ruined. After the parts have soaked for a while and you've rinsed and dried them off, blow all the passages out with compressed air, if you have it available. Be very careful not to get any solvent in your eyes when doing this.

Assembly—Relying on the instruction sheet that came with your rebuild kit, begin assembling your carburetor. First lay out the parts in an orderly fashion. Open up the kit too, but be careful not to lose any of the little parts—ball check, gasket, clips and the like. I leave the parts in the box, but fix it so I can easily see them.

Put the exterior parts on the throttle body first, the economizer/power valve, idle jets and the accelerator pump and its linkage. Be careful during the assembly that you don't over-tighten any component. You are working with delicate parts and soft material. There aren't any torque specifications, so you'll just have to rely on your grey matter.

Start with the inside. Install the main jets using a screwdriver that fits tightly in their slots. The next job is more than just assembly—setting *float level* or *height*. Float level governs the fuel level in the bowl, so it must be right. Check float level by installing the float, its needle and seat. *Lightly* hold the float and its needle in the closed position. Using the float-level gage that came in your rebuild kit, check the float position in the bowl. Position the gage across the top of the float bowl, and measure down to the top of the raised float, using the correct tab on the gage. The kit's specification sheet gives you this information. A tip—rather than holding the float and needle in the closed positions, hold the carburetor upside down and let the float weight do it. This is more realistic as it loads the float assembly similar to how it will be loaded in actual service.

If you find that the float needs adjusting—it will—remove it and its needle assembly to make the adjustment. Bend

Installing the booster-venturi assembly involves several parts. Ball-check and its weight go in first, followed by gasket, booster, booster-venturi screw and its air-distribution plate or gasket.

the end from which the needle hangs, then reinstall it for a recheck. You'll probably have to do this another time or two before the level is right, then you can leave the float and needle permanently installed.

Install the venturi booster assembly, the ball check and its weight, the air-distribution plate and the discharge-nozzle screw to hold the complete assembly in place. You should now be ready to close up your carburetor.

CARBURETOR REBUILD 135

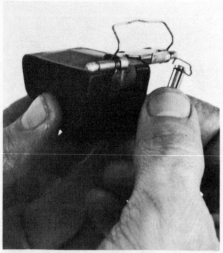

After installing new needle seat, assemble new needle to float and install it in float bowl. Check float height with the gage from the rebuild kit. Gage works for several carburetors, so refer to the specification sheet to find the float-height you are checking. Rather than pushing down on the needle-end of the float to check height like this, hold carburetor upside down so float is hanging.

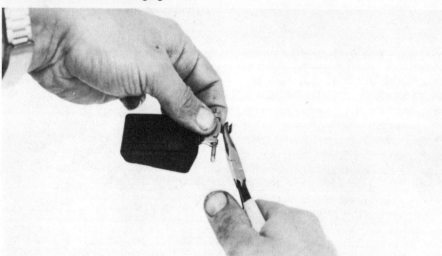

Adjusting float height. Bend down tab from which needle hangs to lower the float. Bend it up to raise float. Install float and recheck its height.

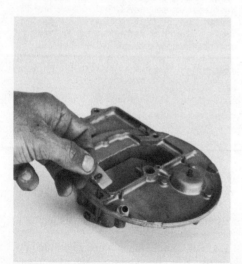

Install choke-linkage dust shield and top gasket. There may be a couple of these gaskets in your kit. Find the correct one by matching it up with your old gasket. Dust shield prevents unfiltered air from being drawn into the carburetor through the choke-linkage clearance hole.

Before installing the bowl cover, install the choke vacuum-break diaphragm assembly if your carburetor is so equipped. A flat nylon seal goes under the cover to seal the diaphragm linkage. Don't forget it. Install the bowl cover with its gasket, four bolts and the center air-cleaner bolt. When installing the idle-mixture screws, run them in until they bottom lightly, then back them off the same number of turns you noted during the teardown. Your carburetor may not be set exactly right for idling, but the idle-mixture adjustment should be close to the correct setting.

To complete the assembly, install all the hang-on parts, the choke, throttle solenoid or dashpot, choke vacuum can and a new filter. Hooking up the linkages to all these parts and getting everything adjusted can turn out to be the most challenging job of rebuilding your carburetor. Follow the kit instruction sheet closely.

TESTING AND CALIBRATION

After your engine is installed with its new or rebuilt carburetor and distributor, you'll want to make sure they will perform at their peaks—maximum mileage and power with a minimum amount of pollutants. Distributor and carburetor tuning are major factors in determining an engine's performance, and to tune them to their full potential requires expensive and sophisticated tuning equipment at the hands of a tuning specialist. This is the subject of the last chapter.

After carburetor top is in place, you can add all the external hardware. Don't forget to replace the identification tag. Finish up by installing throttle solenoid.

9 | Engine Installation

You're now ready for the last big operation, putting your newly rebuilt engine back where it came from—its engine compartment. Get those boxes of bolts, nuts and parts out from where you've been storing them. You'll also need to round up the engine hoist, jack and jack stands again.

Before getting into the installation, thoroughly clean your engine compartment, the accessories, related brackets and hardware so the total job under the hood will look really good. Cleaning an engine compartment requires a spray can of engine cleaner, a stiff-bristle brush for the stubborn grease and dirt and a garden hose. Don't forget to protect your car's body paint. Engine cleaner is strong stuff. You can use the same process to clean all the bolt-on parts and accessories, or you can load them up for a trip to the local car wash. The advantage in doing this is you leave the mess there, and the hot high-pressure water and soap makes the job much easier. Let's get on with installing your engine.

FRONT TRANSMISSION SEAL

Any leakage into the clutch or converter housing which obviously didn't come from the engine should be corrected *now*. This is usually accomplished by replacing the transmission front seal in the case of an automatic, or the bearing retainer gasket and/or seal in the case of a standard transmission.

Automatic Transmission—If you have an automatic transmission, replace the front-pump seal whether it shows signs of leaking or not. Chances are, it will start leaking before you get out of the driveway if you don't, and now's the opportune time.

Before you can replace the front-pump seal, you'll have to remove the converter. Drain the converter before or after it's removed. This will prevent spilling fluid while reinstalling it. It's best to drain it now and fill the transmission after you get everything back together. To drain it before removal, position one of the converter drain plugs at the bottom. Put a bucket or drain pan directly underneath before removing the plug. After the fluid has drained, replace the plug and remove the converter simply by pulling on it, but

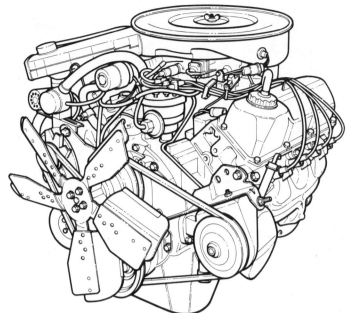

351M/400 lacking only an A/C compressor. Accessory installation can be the most difficult part of an engine installation. Drawing courtesy Ford.

be ready to handle some weight—it's not light. Now, if you decide not to drain the fluid first, pull the converter off and tilt it front-down so the fluid won't run out. You can drain it by tilting it forward.

Removing the converter exposes the front-pump seal. Put a screwdriver behind the seal lip and pry it out. Go around the seal, prying a little at a time and the seal will eventually pop out. The seal can be removed with a chisel and hammer, driving against that little bit which extends beyond the front-pump bore. I don't recommend this because of the risk of damaging the *stator support* or the seal bore. So pry it out.

Before installing the new seal, clean the front-pump bore by wiping it clean of any oil with a paper towel and lacquer thinner. Run a small bead of oil-resistant sealer around the periphery of the new seal, preferably on the edge that engages the pump bore first so the sealer will wipe the full face of the mating surfaces as it is installed. This makes a more effective seal. Just before installing the seal, wipe some clean oil on the seal lip so it will be prelubed. To install the seal, place it squarely in the front-pump bore with the lip pointing toward the transmission and tap lightly around it with a hammer. Be careful it doesn't cock in the bore. Keep doing this until you feel the seal bottom firmly all the way around. Wipe any excess sealer off and you're ready for the converter. While you're still in the engine compartment, clean the inside of the converter housing so it will start out clean. Make sure the front face of the housing—the surface that mates with the engine—is

clean and free of any burrs or nicks. This will ensure a good fit between the engine and the transmission.

Before installing the converter, check the smooth outer surface of the spline, or the surface which runs against the seal, particularly if the seal you replaced leaked. If it has any nicks or burrs or is at all rough, polish it with 400-grit paper. If the surface is deeply grooved, replace the converter. Otherwise you'll end up with a leak just as before. If the surface appears OK, just clean it with a paper towel and lacquer thinner.

Now you're ready to install the converter. Lightly oil the surface you just cleaned. It's added insurance for the seal, even though you already lubricated the seal. It also helps during the converter installation.

Start the converter on the transmission-input shaft and rotate it back-and-forth while pushing on it until it engages with the transmission. Be careful here because the converter must engage the transmission input shaft, but must also key into the front pump and the converter support. As a result you will definitely feel two engagements, and probably all three. When you are certain the converter is on all the way, position the marked flex-plate drive stud at the bottom of the housing in readiness for installing the engine.

Standard Transmission—It's rare when a manual transmission leaks into its bellhousing, but when it does, the culprit is usually the gasket between the front bearing retainer and the transmission case. If your transmission is not leaking, leave it alone because chances are it will never

ENGINE INSTALLATION 137

Replacing automatic-transmission front seal requires removing torque converter. Slide converter off input shaft and drain it.

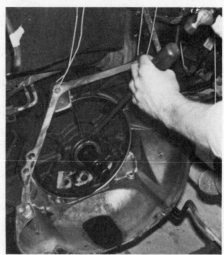
Crude but effecive way of removing front transmission seal—with a screwdriver and hammer. Be careful when doing this so you don't hit stator support or input shaft.

Applying sealer to new seal. Install seal so its lip points toward the transmission. Gradually work it into place by tapping around seal with a hammer and a large-diameter punch. Clean oil and dirt from converter housing.

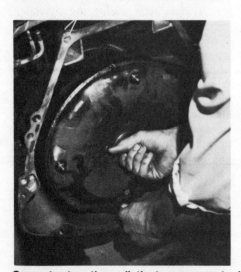
Converter has three distinct engagements, input shaft, stator support and front pump. Make sure you get them all by forcing the converter back while rotating it. You'll be able to feel the engagements. After you're positive the converter is in place, locate the marked drive stud directly at bottom of housing.

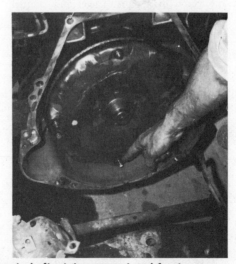

leak. If there are signs of transmission grease in the bellhousing or on the clutch, find where the grease is coming from, the gasket or from around the input shaft. From around the shaft means the input-shaft seal is the bad guy. In either event, the bearing retainer has to come off.

Now that you've removed the bearing retainer, you may as well replace both the gasket and the seal—the gasket will have to be replaced regardless. To remove the seal, pry it out the back of the bearing retainer with a screwdriver. To install the new seal, stand the retainer on its nose on a solid surface, place the seal squarely over its bore so the installed seal lip will point toward the transmission. Carefully drive it into place using a cylinder that matches the seal OD such as a large socket. After the seal is in place, lubricate the lip with oil and reinstall the retainer after you've cleaned it and the transmission-case seal surfaces.

ENGINE BUILDUP AND INSTALLATION

Now turn your attention to the engine and the components which must bolt to it prior to installation.

Clutch & Flywheel—If you have a standard transmission, now's the time to replace or recondition your clutch or flywheel if it's needed. This is purely academic if the clutch slipped before. You'll already know something has to be replaced. The disc for sure, and maybe the pressure plate, or even the flywheel. On the other hand, if it didn't slip, the way to gage clutch wear is by the thickness of your clutch disc. As a disc wears it naturally gets thinner. If not corrected in time this eventually results in pressure-plate and flywheel grooving as the rivets will contact them directly.

Measure the Clutch Disc—To measure a clutch disc, you'll have to compress it to its engaged thickness. When doing this, be careful not to damage the facings or get grease or oil on them. Compress the disc with some sort of clamp. A vise, C-clamp or Vise-Grip pliers will do. Just clamp the disc enough so the wave spring (marcel) is flat. Measure the disc thickness as close as possible to the clamp to get an accurate reading. New disc thickness is about 0.330 inch which allows for 0.050-inch wear, leaving a worn-out disc at 0.280-inch thick. When a disc measures 0.280 inch, the rivet heads are even with the friction surface of the facing and ready to damage the flywheel and pressure plate. As for what thickness you should replace the disc, I don't know except that if it only shows 0.010-inch wear, it obviously

Install standard-transmission front-bearing-retainer seal with lip pointing up as shown. Use a socket matching outer diameter of seal for driving it into place.

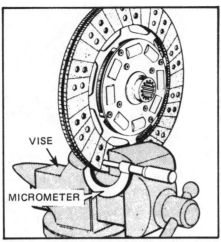

Check clutch-disc thickness by clamping it just enough to fully compress it. If it measures near 0.280 inch consider it worn out and replace it. Drawing courtesy Schiefer.

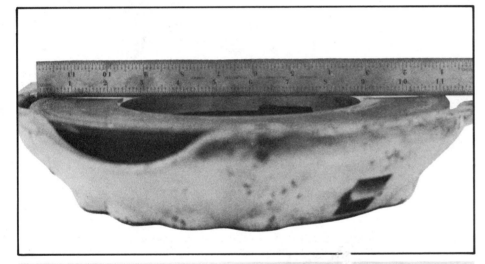

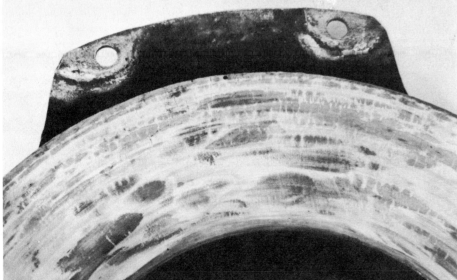

Pressure plates show effects of overheating. One is warped as can be seen by the gap between the ruler and pressure plate. Heat checks and chatter marks or hot spots on bottom pressure plate indicates it has been extremely hot. If your pressure plate or flywheel is suffering from either of these symptoms, replace it.

doesn't need replacing. However, if it's down to 0.290 inch, replace it. Anything in between depends on how long you want to drive before you have to replace the clutch disc.

Pressure-Plate Inspection—Assuming you have the disc problem sorted out, inspect your pressure plate. The general rule is, if the disc wore out in a short time from one thing or another, the pressure plate may require replacing, but if the disc looks good or gave you many miles of good service, chances are it's all right. The first thing to do when deciding whether a pressure plate is reusable is to simply look at it—give it a visual inspection. If the friction surface is bright and shiny, and free of deep grooves, the pressure plate is probably OK. However, if it has black and blue spots (hot spots) covering it, it has been slipped excessively and may be full of *heat-checks* (cracks). In any case, check the friction surface by laying a straight-edge across it. If is is warped, it will show an increasing air gap between the straight edge, starting from nothing at the outside edge and increasing toward the inside edge—it will be *concave*.

If the hot spots, heat checks or warpage don't look too bad, you should compare the price of resurfacing your pressure plate versus the price of a new or rebuilt one if your budget is tight. Otherwise, don't waste time—replace it. Regardless, if your clutch has survived more than two discs it should be replaced no matter how good the friction surface looks. The bushings in the release mechanism will be worn and the springs which apply pressure will be weak from fatigue, resulting in reduced clutch capacity. You may have gotten away with it with a tired engine, but now your engine will be making more torque.

Flywheel Inspection—Because the flywheel *sees* the same slippage as the pressure plate, it will suffer from similar ills. If your pressure plate has problems, take a really good look at your flywheel too. Because it has more mass than the pressure plate pressure ring, it will absorb more heat before being damaged. For instance, you won't see a warped flywheel because the heat required to do it would destroy the friction facings first. On the other hand, it is quite common for a flywheel to get hot enough to heat-check. Also, a flywheel is not immune to rivet damage, it can be grooved as easily as a pressure plate. So, if your pressure plate is grooved, chances are the flywheel is too.

If your flywheel is heat-checked or badly grooved, it should be resurfaced.

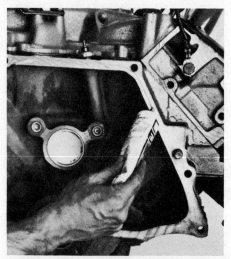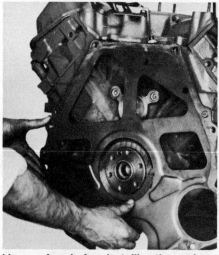

I apply adhesive-type gasket sealer on the block's rear-face before installing the engine plate. This keeps it from falling off its dowel pins and down between the engine and flywheel or flexplate at the most inopportune time—during installation.

It's possible that surfacing won't correct a seriously heat-checked surface and the flywheel will have to be junked. Only so much material can be removed from a flywheel before it becomes dangerous. And reusing a heat-checked flywheel is equally dangerous. Heat-checks can develop into long, deep radial cracks (like the spokes of a wheel) toward the center of the flywheel. If this does or has happened, the flywheel can't be repaired by resurfacing. More important, it's dangerous! A cracked flywheel can easily fly apart or explode with the force of a hand grenade.

To have your flywheel resurfaced, take it to your engine machine shop. Make sure the job is done on a grinder rather than a lathe, particularly if yours is hot-spotted. They are very hard, consequently if the flywheel is resurfaced by machining, the cutting tool will *jump* over the hard

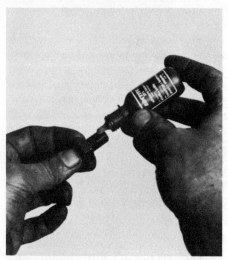

After rotating flywheel or flexplate to line up with crankshaft mounting holes, Loctite bolts and install them. Tighten and torque bolts 75—85 ft.lbs. in a criss-cross pattern. Some flexplates use a load-spreader ring. If yours does, install it under the mounting bolts. A screwdriver in a flexplate hole keeps the crank from rotating as you torque the bolts.

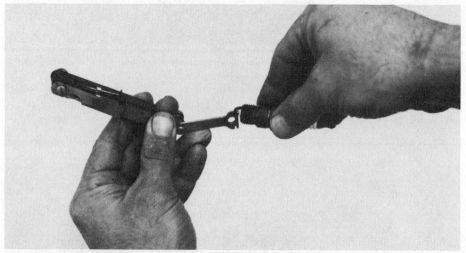

Old transmission-input shaft aligns disc with pilot bearing. Tighten each pressure-plate bolt a little at a time, then remove input shaft.

Even though you're installing new plugs, check and gap them. I've seen the gaps closed up on new plugs, probably from the box being dropped. Tapered-seat plugs are correctly tightened by giving them 1/16 turn after they are finger-tight. This is equivalent to 7—15 ft.lbs. for 14mm plugs and 15—20 ft.lbs. for 18mm plugs.

spots, creating little bumps on the friction surface. A grinding wheel cuts right through the hard spots.

Although your flywheel appears to be all right, it should still be cleaned. For instance, the clutch friction surface will be shiny, or may have some deposits on it. To help it and the clutch, use fine-grit sandpaper to roughen the surface. This will remove resin deposits left from the clutch disc too. Follow the sanding with a good cleaning with non-petroleum-based solvent such as alcohol or lacquer thinner. This will remove any oil deposits. Give the same treatment to the pressure plate if you are planning to reuse your old one.

Clutch Release Bearing—There are two rules when it comes to the release bearing: replace it if you are replacing your clutch. The next rule is replace it anyway. The relative cost of a new bearing and the amount of work it takes to replace it now as opposed to when your engine is back in should give you a good idea as to what you should do. Regardless of whether or not you replace your bearing, fill the bearing-hub-bore groove with grease. Don't overdo it because too much grease causes clutch failure if the grease gets on the clutch disc. Use moly grease to lubricate the bearing hub and the wear points on the release lever, then replace the bearing.

Engine Plate—Whether you have an automatic transmission or a standard transmission, there's an engine plate to install before the flywheel/flexplate can go on. Because it depends on the bellhousing or converter housing to hold it to the engine, it'll be constantly falling off while you are trying to install the flywheel/flexplate. It can fall away from the block and down between the block and flywheel/flexplate while you're trying to bolt the engine to the transmission. Here's a tip: Apply a thin bead of 3M weatherstrip adhesive to the rear face of your engine block to hold the engine plate in place. Install the plate over the locating dowels with the starter hole to the right. Hold it in place for a minute or so to make sure the adhesive gets a grip on the plate.

Flywheel or Flexplate—Make sure your crankshaft-mounting flange is clean and free of burrs, particularly in the case where you will be mounting a flywheel to it. This can cock the flywheel, causing it to wobble, which will lead to clutch and transmission problems.

Install the flywheel or flexplate using the 6 grade-8 bolts and a drop of Loctite on the threads. When locating the flywheel or flexplate, rotate it on the crank-mounting flange until the mounting holes line up. They are drilled so the flywheel/flexplate can mount in only one position so the engine will be balanced correctly. Don't panic when you see the holes don't line up the first time. Snug the mounting bolts, then torque them to 75–85 ft.lbs. Position the crankshaft so the *marked* converter-mounting hole is at the bottom.

Mounting the Clutch—As soon as the flywheel is mounted on the engine, it's ready for the clutch. Remember to avoid touching the friction surfaces—pressure plate, flywheel and disc. Grease on the clutch surfaces causes bad things like grabbing or slipping, so be careful not to touch the operating surfaces.

To mount the clutch, you'll need a tool to align the clutch disc with the center of the crankshaft while you are tightening down the pressure plate. An old transmission input shaft works well, or special inexpensive tools are available for doing this job. Clutch-alignment tools are available at most automotive parts stores.

Although you can probably handle this job yourself, a friendly third hand comes in handy. To install the clutch, hold the pressure plate and the disc against the flywheel while starting at least two pressure-plate bolts and their lockwashers. Make sure you have the disc installed in the right direction. FLYWHEEL SIDE should be indicated on the disc, however if it's not, the spring-and-hub assembly goes toward the transmission.

With the bolts holding the pressure plate up, install the input shaft or clutch-alignment tool in the clutch disc and center it in the pilot bearing. You can now start the rest of the bolts. A word of caution—*don't tighten any of the bolts all the way*. This will result in junking the pressure plate because it will bend the cover. Tighten each bolt a couple of turns at a time. Go around the pressure-plate bolts several times until the cover is firmly against the flywheel. Remove the alignment tool and torque the pressure-plate bolts to 12–20 ft.lbs.

Spark Plugs—If you haven't already done so, install a new set of spark plugs. They will be well protected in the heads, particularly with the exhaust manifolds installed. Remove the old plugs you installed to keep the cylinders clean and dry, and replace them with new ones. Check and set their gaps first. Torque all the plugs to 10 ft.lbs. except for the one in number-1 cylinder. Leave it loose.

Engine Mounts—Engine-mount designs vary according to the engine and chassis they are used with, however their basic design does not change. All use a single through-bolt for ease of engine installation. Some mounts are not interchangeable from side to side, so make sure you identify which is which. R.H. or L.H. should be stamped on the mounts somewhere if they don't interchange. Also, I'm sure you remember that some installations require that the engine-mount insulators be removed so the engine can be removed and re-installed. This is the case with 460 truck installations due to the tight fit between the firewall and the transmission bellhousing or converter housing. Once the engine and transmission are connected, you can re-install the insulators.

INSTALLING THE ENGINE

There's not much more you can do with the engine out, so the big moment has come. Attach a chain or cable to the front of the left head and the rear of the right head or, better yet, make use of the lifting lugs at the exhaust manifolds. Now, if you have an automatic transmission, make sure the marked hole on the flexplate is at the bottom. The marked stud on the torque converter goes at the bottom too. It won't take long to find out how important this is if the converter won't engage the flexplate.

Lift the engine and check to see that it's level side-to-side and slightly low at the rear. Set the engine down if necessary and adjust the chain until it's right. This

The moment we've been waiting for, installation time. While Louis controls the engine, notice flexplate mark positioned at bottom, ground strap attached to the right head, all loose ends are up out of the way and wiring harness is installed. Engine can be lowered while care is taken to make sure everything clears. You can't see it, but there's a jack under the transmission to bring it up to where it will engage with the engine.

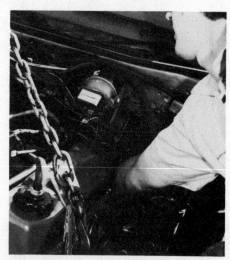

Object here is to engage converter drive lugs with the flexplate or standard-transmission input shaft with the clutch and the engine dowels with the transmission housing, then to get one bolt started. It sometimes takes a lot of wrenching and joggling until this task is accomplished, but let's not give up now.

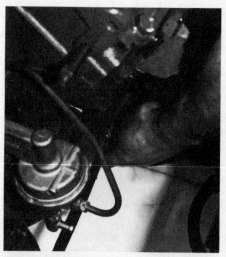

With engine and transmission engaged, lower engine until holes in one mount line up. Slide through-bolt into place, then install other one. Nuts can be installed once you have your car up in the air.

will save a lot of unnecessary trouble later when you're trying to mate the engine to the transmission. Position a floor jack under the transmission and raise it up enough so you can remove the wires or whatever you used to support the transmission and exhaust pipes. If you have a manual transmission, *lightly oil* the crankshaft pilot bearing and put a light film of moly grease on the transmission input shaft and bearing retainer. Put the transmission in gear. Look over the engine compartment to make sure everything is tied up out of the way so nothing will interfere as the engine is being lowered into place.

Lower the Engine In—Position the engine over its compartment and gently lower it while making sure it doesn't hang up on something. Jack up the front of the transmission as high as possible so the engine can be lined up easily and engaged with the transmission while clearing the engine mounts. This is particularly true with a manual transmission. For a manual transmission, carefully guide the engine back, centering the clutch disc on the transmission-input shaft. When the disc and input-shaft splines contact each other, rotate the engine back-and-forth while holding it against the input shaft until the splines engage. Then put the transmission in neutral so you can rotate the engine into engagement with the bellhousing. This means aligning the engine dowel pins with the bellhousing holes. When this happens, the two will literally "click" together, signaling you to get at least one bolt in place to hold the two together.

With automatic transmissions, the converter studs and the engine dowels must engage almost simultaneously. Proceed as with a manual transmission and push on the engine while rocking it, but not as much. If you have things lined up well, this job will only be difficult. You'll have to climb right in there with the engine and the converter housing. This is so you can see the relative positions of the two. When the two come together, get some bolts in to keep them together. Also check to see that the converter is engaged with the flexplate.

Engine-Mount—With engine and transmission mated, you can remove the jack from under the transmission—unless you have a truck. You still have the insulators to install between the frame and engine-mount brackets. Remember, there could be a right- and a left-hand insulator, so make sure you're installing them in the right positions. It's tough enough to put them in once, let alone getting them in wrong, then having to take them back out and switch sides.

The first insulator is the easy one. The engine rocks over enough so it'll slide right into place. Once installed though, it keeps the engine from rocking the other way far enough to make room for installing the last one. Consequently, you'll have to lift and pry the engine up while being careful not to damage anything before you can juggle the last insulator into place. All I can say is, it's a tough job! Ford designed the mounts for installing the engine and tranmission as an assembly, not this way. The law that says things come apart easier then they go back together applies doubly here. Get the vertical bolts into place and you can continue.

Lower the engine while lining up the engine-mount holes, then slide the horizontal bolts into place. Get the nuts on the bolts unless it would be easier from above. Lower the engine completely and you can remove the hoist—you won't need it anymore. Jack your car up and set it on jack-stands. With the car *firmly* supported, slide underneath and finish installing the engine-mount through-bolts. A box-end wrench on the head of the bolt to keep it from turning and a long extension with a socket and ratchet at the other end is the easy way of tightening them.

Bellhousing/Converter Housing to Engine—Install as many bellhousing or converter-housing bolts as you can manage from underneath. Torque them to 40 ft.lbs.

Starter Motor—Make sure when you're gathering the starter motor and its bolts that you identify the right bolt for the bottom. This bolt is special and can be recognized by its reduced 1/2-inch hex size and the integral washer forged into the head. A normal 7/16-inch bolt has a 5/8-inch hex. The reason for this special bolt is to provide wrench clearance between the starter motor and the bolt head. The top bolt threads into the starter from the back side of the bellhousing rather than into the bellhousing from the front.

When putting the starter in place, you'll find that it is best done by feeding the front end up over the steering linkage—in the case of rear-steer passenger cars—then back it into its opening in the bellhousing. Once you have it in place, install a bolt to hold it there. Run the other bolt

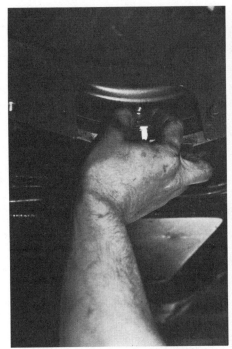

With the converter-housing installed, complete your engine-to-transmission hookup by installing converter-to-flexplate nuts. Wrench on crank damper is for turning crankshaft to expose converter studs.

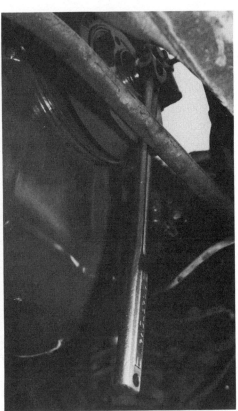

Seal converter housing after you have converter firmly attached to flexplate. Don't forget engine-plate-to-converter-housing bolts.

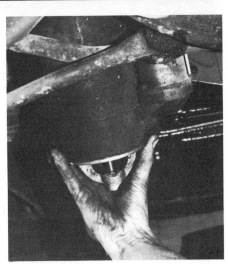

Get front of the starter up in there first, then slide its gear end back into converter or bellhousing.

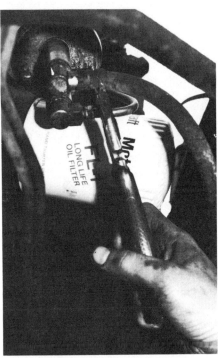

Install oil filter and connect fuel line to fuel pump while you're still down below. Hose clamp finishes this job.

in and torque them to a maximum of 20 ft.lbs.

After you have the starter motor in place, connect its battery lead if disconnected. Route the cable through any clips or brackets it may run through on its way to the battery.

Battery and Alternator—If you want to use the starter to turn the crank for fastening down the torque converter to the flexplate, you'll have to install the battery, connect it to the starter-solenoid lead and connect the engine ground cable to the engine and the battery. Because the battery ground and the alternator leads connect to the engine, hang the alternator on the engine loosely. Don't forget the spacer if one is used. Connect the two leads to the block *with the star washer against the block* to ensure a good ground. With a friend at the ignition switch or with a remote starter switch, you can now turn the crank the modern way. Otherwise, you can simply rotate the crank using a socket and handle on the crankshaft-damper bolt. This last method is easy with the spark plugs out.

Converter-to-Flexplate Nuts—The converter can now be firmly attached to the flexplate. Remember, you'll have to tighten one nut, then rotate the crank 90° to get to the next one until you have nuts on all the studs. Be careful if you are using a friend at the ignition switch. Make sure he understands the switch shouldn't be touched *until you say so*. Torque the nuts to 25 ft. lbs.

Engine-Plate to Bellhousing—The engine plate extends down in front of the bellhousing/converter housing. Thread its attaching bolts into the bellhousing and torque them 25 ft.lbs.

Exhaust Pipes to Manifolds—Connect the exhaust system now. If yours has replaceable exhaust-pipe-to-manifold gaskets, you'll find them in the gasket set. They are the conical-shaped rings approximately 3 inches in diameter.

ENGINE INSTALLATION 143

Installing alternator with its two brackets and spacers. Don't tighten pivot bolt or one at the adjusting bracket yet.

Try doing this without pouring oil on your clean engine. If you don't have a spout, make sure there's a rag handy. Regardless of the brand of oil you're using, just make sure it's SE grade.

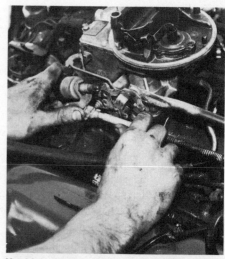

Hooking up this 400's accelerator cable. The automatic-transmission kick-down rod has already been installed. Next comes the cruise-control linkage.

Pilot the exhaust pipes into the manifolds, slide the mounting flanges up each pipe, then hold them in place with a nut on one of the studs. This will support the pipe too. Run the other nuts on and torque them all to 30 ft.lbs.

Oil Filter—Check before lowering the car. It may be easier to install the oil filter from below. Fill it about half full first so you'll get oil pressure sooner. Smear some oil on the seal, then run the filter up until it contacts the block, and give it another half turn. While you're under there, check the oil-pan drain plug just to make certain it's there *and tight*. Set the car down now that you're ready to go top-side.

Engine Oil—If for no other reason than to take a break, now is a good time to fill the crankcase with oil. Because the filter has some oil in it, add a quart less than what's specified. You can check oil level after running the engine and add oil as necessary to bring it up to the mark.

Carburetor Linkage—The accelerator linkage will be either the rod-and-lever or cable type.

If you have a cable type, there will be a bracket which attaches to the intake manifold to mount the carburetor end of the cable conduit. Install it if it's not already on the manifold. The cable will either clip or bolt onto this bracket. At the carburetor end of the cable, there will be a socket which attaches to a ball on the carburetor bellcrank. Shove it on with your thumb and it'll stay in place.

The rod-type linkage attaches the same way, but it doesn't require a manifold bracket. The rod will extend from a lever mounted on the firewall. This type is used on most trucks.

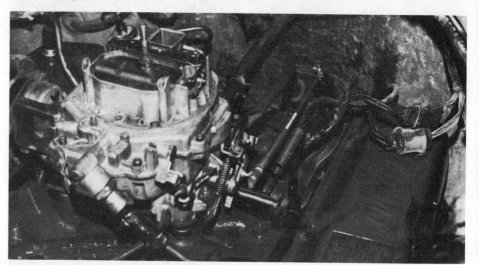

Carburetor linkage on 460-powered truck uses a rod-and-lever system rather than a cable linkage. Note bracket used to stabilize automatic-transmission kickdown rod. It must be used, otherwise rod will flop from side to side.

The next item you'll have to deal with at the carburetor is the automatic transmission TV, or kick-down rod. Slip it over a pin extending from the movable lever on the carburetor bellcrank. There's a special clip which holds this link in place. It clips in a groove in the pin so it is spring-loaded against the side of the rod. Make sure you have all the return springs in place. There'll be one for the TV rod and two for the carburetor return. Federal mandates require the auto makers to use two throttle-return springs at the carburetor.

Ignition Coil—Over the years, you could find an ignition coil mounted just about anywhere. However, Ford seems to have settled on mounting most coils on the top-right of the intake manifold. This is a good spot because nothing else seems to occupy the space but, better yet, it's kept cool due to being in the air blast from the fan. Locate the coil bracket to the notch and hole made for it, and secure it with its one bolt.

Wiring Harness—Locate the engine-wiring harness along the inboard side of the left valve cover. Retain the harness using the clips you've installed under the valve-cover bolts. Some of the leads can be connected now and some can't. For instance, you should be able to connect the sending-unit leads for oil pressure and water temperature, and the leads to the coil. The distributor and A/C-compressor clutch leads will have to wait.

Fuel Line—Connect the fuel line from the fuel tank to the pump. Also, install the fuel-pump-to-carburetor line if you haven't done so yet.

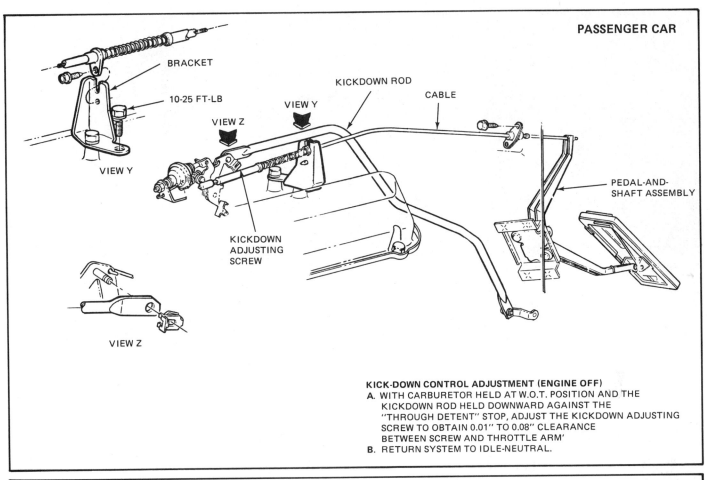

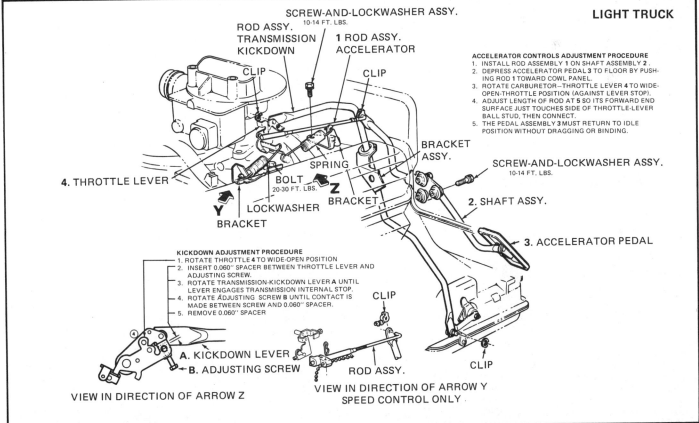

Typical passenger-car cable linkage and truck rod-and-lever linkage. You'll find at least two throttle-return springs used with both types of linkages. Drawings courtesy Ford.

ENGINE INSTALLATION 145

High-light damper timing mark with white paint before installing and statically timing distributor. Remove number-1 plug and use "finger-over-the-hole" technique to find TDC. Back crank up to timing mark, then install your distributor.

Approximate position your rotor and vacuum advance/retard can should be in after your distributor is properly installed. Rotor turns counter-clockwise viewed from above, so rotating the housing clockwise advances timing and counter-clockwise retards it.

Some moly on the gear, a new O-ring at the base of the housing and a rubber band to hold the distributor-cap retaining clips up out of the way and this distributor is ready to be installed. Notice number-1 cylinder mark on the housing. Rotor should be pointing to it when distributor is in correct position.

Install distributor by starting with the vacuum can pointing forward and the rotor pointing rearward. Rotor will turn clockwise as distributor gear engages drive gear. If you're lucky, distributor will also engage oil-pump driveshaft, but this is unlikely. Push distributor down like this while cranking the engine or rotating the crankshaft back-and-forth until it does engage, then bring the crank and rotor back to number-1 firing position.

Number-1 Cylinder on TDC—The distributor will be installed so it is in position to fire number-1 cylinder. This requires that number-1 cylinder be set at TDC of its *power stroke*. Remember, the number-1 TDC mark on the crankshaft damper passes the timing pointer twice for each complete cycle, once between the exhaust and intake strokes and once between the compression and power strokes. The latter is the one you want.

Of the many ways to find TDC, I'll stick to the easiest one. It doesn't require using much more than common sense. Start by removing the plug from number-1 cylinder—that's why I suggested leaving it loose. Put your thumb over the hole while cranking the engine. When you feel the compression against your thumb, stop cranking. You'll know the piston is coming up on its compression stroke. All you have to do now is rotate the damper clockwise to its TDC mark. Better yet, set the mark on the initial BTDC position—about 10°—and you're ready to install the distributor. Reinstall the spark plug, but tighten it this time.

Installing the Distributor—With your engine *timed*, you can now install the distributor. To time the distributor with the engine, you need a way to set the rotor in position to fire cylinder 1. Install the distributor cap as a reference and mark the side of the distributor housing with grease pencil or tape in line with the distributor-cap socket 1. Then remove the cap. This will allow you to verify that the rotor is in position to fire the number-1 spark plug.

Before installing the distributor, double-check to make sure the O-ring is installed on the base of the distributor housing, then oil it. As you drop the distributor into place, keep the vacuum-advance can pointed forward and the rotor pointed rearward. The tip of the rotor will be approximately an inch from the number-1 mark on the distributor body. As the distributor-drive gear engages the camshaft gear, the rotor will turn clockwise bringing the rotor in line with the mark, providing you keep the vacuum can pointing forward. If the rotor doesn't line up with the mark, pull the distributor out for enough so you can jump a tooth in the direction it must go to line up. A little care here will ensure that correct timing can be achieved without positioning the distributor housing at an awkward angle.

I've made this procedure sound easier than it really is. Not only must the distrib-

Statically timing the distributor. With crank on 10° BTDC, rotate housing clockwise so rotor lines up with the number-1 mark and points *just begin* to open.

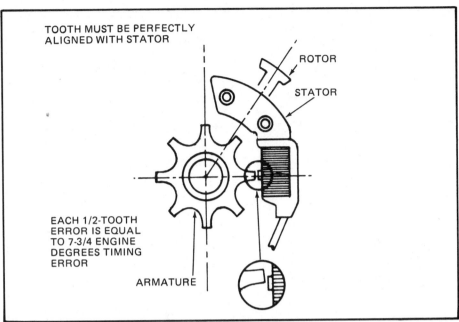

Statically time a solid-state distributor by rotating the housing so one armature vane lines up with the magnetic pickup as shown in the circle. Rotor should also be located midway between two magnetic-pickup rivets. Drawing courtesy Ford.

utor gear engage its drive gear on the camshaft, the distributor shaft must also engage the oil-pump drive shaft. The chance of these two engaging on the first try is highly unlikely. Once the distributor gears have partially engaged and the distributor has ceased to go into the block any farther, bump the engine over. This will rotate the distributor shaft so it will eventually line up with the oil-pump shaft. You won't have to rotate the crank very far. Apply a slight downward pressure on the distributor until it engages, then bring the crank back to TDC and check the rotor position.

With the distributor in place, you can come fairly close to correct timing before the engine is run by timing it *statically*. To do this, turn the crank so the pointer and crank damper indicate the correct distributor advance—10° BTDC for example. Install the distributor hold-down clamp and bolt loosely, and then rotate the distributor housing clockwise, against the rotation of the rotor, until the points *just* begin to open. Spark occurs the instant the points *break contact*. Secure the hold-down clamp. Later, when you get the engine running, check the timing with a timing light and readjust as necessary.

If your distributor is the later breakerless type, a similar method works. However, rather than having the points to go by, align one of the armature spokes as shown in the sketch. Looking down on the distributor, the rotor should be between the two rivets which secure the magnetic pickup when it is in the number-1 firing position.

ACCESSORY-DRIVE INSTALLATION

The difficulty of installing engine accessories depends simply on how many your engine has. It can be a five-minute job or the toughest part of the engine installation. I'll treat the installation assuming your car or truck is loaded and you can skip those sections that don't apply.

A/C Compressor and Power-Steering Pump—If an engine is equipped with both power steering and A/C accessories, they have to be installed together because they share common mountings. *Hang-on-A/C* units are a different matter—they will be separate.

The A/C compressor should be hanging right where you left it—on the left-front fender apron. If you removed it, install its bottom mounting bracket. Make sure you use the correct mounting bolts if you have the two-piston-type compressor. They are the short 3/8-16 bolts with integral serrated-face washers. Ford terms these *UBS bolts,* meaning "uniform bearing strength." Make sure you use the correct length bolts and install them carefully to check for "bottoming." The very short ones are used to mount the adjustable-idler bracket to the compressor bracket. Torque these bolts 20—30 ft.lbs.

If they aren't already there, install the A/C-compressor and power-steering-pump mounting studs in the left cylinder head. loosely install the A/C compressor and its bracket on the studs, then install the power steering over the A/C bracket. There'll also be a stud at the water pump for mounting the inboard end of the power-steering-pump bracket. Run some nuts on the studs to hold everything in place. If you have a back-side idler for the A/C belt, it mounts on top of the A/C compressor and power-steering pump brackets. Tighten everything securely once you get all the brackets fitted and figured out.

Turn your attention to the top A/C-compressor bracket. It will fit between the compressor and the water pump. It not only helps support the compressor,

ENGINE INSTALLATION **147**

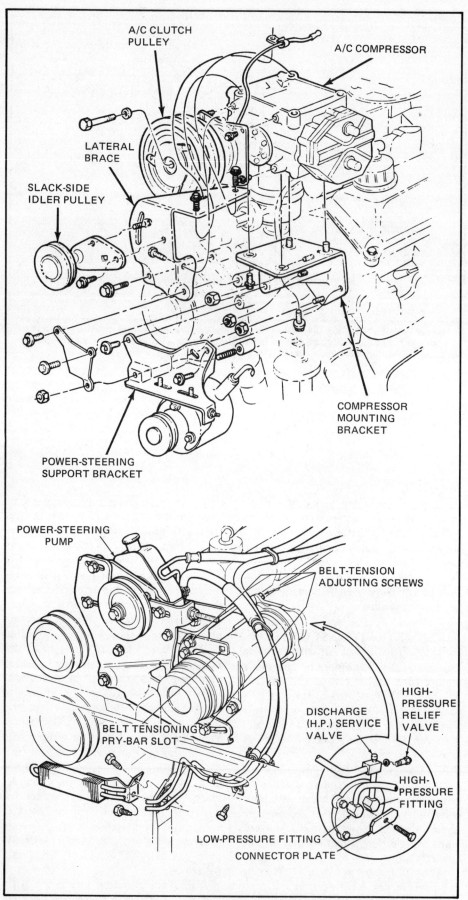

Typical A/C-compressor and power-steering pump installations. You'll find some six-cylinder compressors mounted on top rather than below as in the case with this 429/460 installation. Drawings courtesy Ford.

it is also used to mount the adjustable idler pulley for the compressor drive belt and the high-pressure A/C line, power-steering hose and the retaining clip for the compressor-clutch wire depending on your vehicle. UBS bolts are also used here, two at the top and one in the backside of the compressor. Torque the bolts 20—30 ft.lbs. while making certain they are short enough so they pull down on the bracket rather than bottoming in the compressor housing. If it's not already on the bracket, install the idler pulley, but don't tighten it yet.

Rotary compressors have six pistons operated by a swash, or wobble plate. They mount similar to the two-piston type in that they share brackets with the power-steering pump. Other than having totally different bracketry, the major difference is that the six-piston compressor may be mounted above or below the power-steering pump, depending on its application, whereas the two-cylinder compressor will always be on the top.

A/C Compressor Wiring—To finish the accessory installation job on the left side of your engine, connect the compressor's magnetic-clutch lead. Do it now because it's easy to forget.

Air-Injection Pump and Alternator—If your engine is equipped with an alternator and an air pump, they'll use common brackets. Because you may already have the alternator loosely mounted, all that's left is the air pump if your engine is so equipped. Just as you did with the A/C and power steering, mount them loosely until you're sure you have all the spacers, bolts and brackets in their right locations, then you can tighten them up.

Route the air-pump hoses and mount the bypass valve. Make sure the air hoses are in good condition by squeezing them. If cracks appear, they are in bad shape and should be replaced.

Spark Plug and Coil Leads—Turn your attention back to the ignition system. Make sure the rotor is pushed all the way down on the distributor shaft. Install the distributor cap and secure it with the two spring clips. If they're not already in place, install the little plastic ignition-wire retaining clips. Starting with socket 1 on the distributor cap, route the wires in the engine firing order as you go around the cap counter-clockwise, point the wires in the general direction of their valve-cover clips. Remember, the firing orders are: 1-5-4-2-6-3-7-8 for the 351C and 351M/400 and 1-3-7-2-6-5-4-8 for the 429/460. The order for the wires in the right valve-cover clip is 1-2-3-4 from front to back.

Object here is to get that first bolt into place. A/C compressors are heavy.

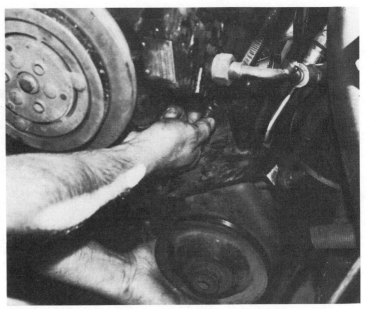
With the A/C compressor loosely installed, the power-steering pump can go on. They use many common mounting bolts, consequently they must be installed together.

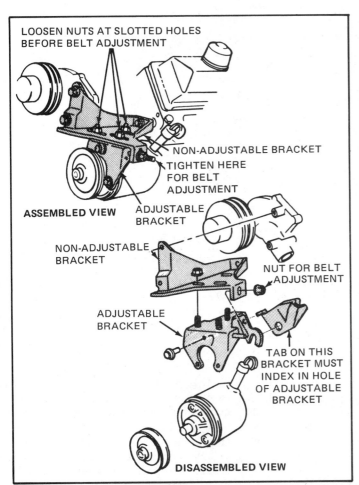
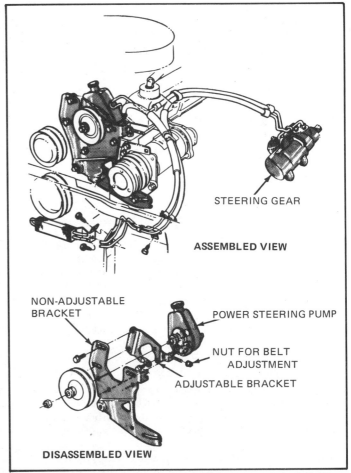
Ford power-steering-pump installation on a 351M/400 and Saginaw power-steering-pump/low-mounted-A/C-compressor installation on a 429/460. In both instances, pumps are mounted to brackets mounted to another bracket for belt adjustment. Drawings courtesy Ford.

ENGINE INSTALLATION 149

Installing top A/C compressor bracket and adjustable idler pulley. Use UBS-style bolts of the correct length and don't miss one hidden behind bracket.

Working the power-steering-fluid cooler into place between the A/C compressor and the rigid suction line. It's not easy and is worth a few scraped knuckles, but it beats disconnecting an A/C line and losing the refrigerant.

An important little item, the spark-plug-wire clip. This keeps the wires right where you want them, off the engine and neatly routed.

Because two cylinders fire one right after the other in the left cylinder bank, the ignition wires are not installed in order in the clip. Order for the 351C and 351M/400 engines is 7-5-6-8 and for the 429/460 the order is 5-7-6-8. This prevents a possible misfire as a result of *induced current* in wire number 8 in the 351C and 351M/400 and number 5 in the 429/460. An induced current occurs when two wires are close and run parallel for some distance. Current flowing in one of the wires automatically *induces* current to flow in the other wire, enough to cause a weak spark and a misfire.

Spark-plug wires have different lengths. Consequently, the easiest way to install new wires, if you're doing so, is to install the old distributor cap and wires loosely. Now, all you have to do is replace each wire one-at-a-time by comparing the lengths and duplicating the routings. However, if you've discarded your old wires, you'll have to do a trial routing. Install the wires loosely at the distributor cap. They are difficult to remove if they are fully installed. After you are satisfied with the routing, push down on the molded cap at wire end to seat the terminal in its socket. It's not easy because, as the cap goes on, air has to be forced out. At the spark-plug end, push until you feel the connector click into place, then give it a little tug to see if it's on all the way. Pull on the molded sleeve, not on the wire. Don't forget the coil lead. It's easy to install once the plug wires are in place.

Heater Hoses—Install new heater hoses as you would a radiator hose. If they are over two-years old, replace them and use the old hoses to determine the length of the new ones. The heater-inlet hose is routed from the intake manifold on the 429/460, or the block on other engines, to the *bottom* heater-core tube. This means the heater-outlet hose is routed from the top heater-core tube to the water-pump inlet. Route the heater-outlet hose through the automatic-choke clip on models that have one.

Miscellaneous Hoses—The last major job is to route the vacuum hoses. Again, the complexity of this depends on the type and number of accessories and the emissions devices used. Here's where your labeling of the hoses and photos you may have taken will be invaluable. Before you begin installing the hoses, check them for signs of cracking. If some have begun to crack, replace all of them. A cracked vacuum hose can be very hard to find and it really complicates getting your engine running right.

Depending on your particular vehicle,

Route spark-plug wires according to your engine's firing order beginning with the number-1 plug: 1-5-4-2-6-3-7-8 for 429/460s and 1-3-7-2-6-5-4-8 for the 351C and 351M/400. Note number-5 and -6 leads (arrows) are separated on this 351C as they should also be on a 351M/400. This prevents induction misfiring of number-5 plug. For the 429/460, separate number-7 and -8 leads in similar manner.

If your choke has a clip for routing the heater hose like so, run the hose which goes from the 429/460 intake manifold or 351C and 351M/400 block to the heater-core inlet through clip.

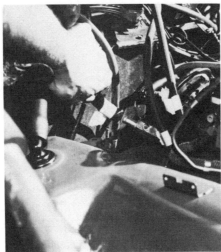
Time spent labeling all those wires and hoses pays off now. Hooking up vacuum hoses and wiring at back of engine.

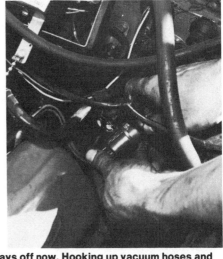

Inspect the fan carefully for cracks. After positioning water-pump pulley so its mounting holes line up with the water-pump drive flange, install the loose-assembled fan, spacer and four bolts.

Regardless of the type fan, you'll have to keep the water-pump shaft from turning as you tighten the fan bolts. Simply hold one blade of a fixed- or flex-blade fan.

For the viscous-drive fan, use a screwdriver behind the head of a bolt and the fan shaft to keep it from turning when tightening its bolts..

vacuum hoses you may have from various vacuum taps at the back of the intake manifold are for: power-brake booster, automatic-transmission throttle valve, heater and air conditioning vacuum motors, the distributor-advance and emissions devices such as an EGR valve with its hose routed through a PVS (ported vacuum switch) in the thermostat housing. The PVS meters engine vacuum to the distributor and EGR valve in some instances, but not all. The various types of emissions equipment and the vacuum circuitry are beyond the scope of this book, so the correct routing of your system depends on how well you labeled or photographed your engine during the removal process.

Ground Strap and Transmission Filler Tube—Because of their similar locations and difficult access, install the engine ground strap and the transmission filler tube to the back of the right cylinder head. In some cases the ground strap may be attached to the intake manifold. When attaching the ground strap, make sure you install a *star washer* between the head or manifold and the terminal. The washer ensures a good ground between the connector and the cylinder head, particularly in geographic areas where corrosion is a problem.

Water-Pump Pulley and Fan—In preparation for installing the radiator, install the water-pump pulley and the fan. To make the job easier, locate the pulley on the

ENGINE INSTALLATION 151

Time for the belts. Make sure they're new or in great shape. Loosely install them from back to front.

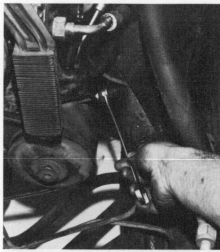

Tightening belts the easy way—with a breaker bar at the A/C compressor and an adjusting nut at the power-steering pump. Readjust the belts for the proper 1/2-inch deflection after you've run your engine.

water-pump shaft and line up the holes in the pulley with those in the water-pump flange *after* loosely installing the alternator and air-pump belts if you have a 429/460. This is necessary because there's not enough room between the water-pump and crankshaft pulleys to slip the belts on their pulleys. If you have a fixed fan, loosely assemble the fan, spacer, the four mounting bolts and lock washers and locate this assembly on the water-pump shaft and start the bolts. Tighten the bolts in a criss-cross pattern and torque them 10—15 ft.lbs. You'll have to hold the fan to keep it from turning while tightening the bolts.

If your fan is the flex-blade type, inspect the blades for cracks. Because of the constant flexing, the roots of the blades can eventually fatigue to the point of cracking and breaking off. As a result, this type of fan has been known to throw a blade when the engine is running, so remember, *never stand in line with a fan when the engine is running,* particularly when the engine is operating faster than idle.

When installing a clutch-drive fan, there's no spacer, but the job may still be difficult because of the way the bolts are located behind the fan and clutch-drive assembly. The fan is not fixed, so you'll have to wait until the drive belts are installed before you can torque the mounting bolts. So for now, tighten them as much as you can, then complete the job after installing and adjusting the belts.

Accessory-Drive Belts—Your engine will be equipped with one to four belts, again depending on the accessories your engine is equipped with. If they are more than two-years old, replace them. Inspect them regardless. If you have more than one drive belt, install them in the proper order back to front. You should have already started with the alternator or air-pump belt if you have a 429/460.

It's very important that the belts be adjusted to their proper tension. If one is too loose, it will slip on the pulley. If the belt happens to be driving the alternator, you'll be plagued by a battery that seems to run down for no reason. Conversely, if a belt is too tight, it will overload all the bearings in the pulleys and shafts it drives and is driven by. This is particularly important in the case of a water pump. They don't take kindly to being overloaded. It shortens a water pump's bearing life drastically.

As for how tight a drive belt should be, Ford specifies 140 lbs. tension in a new belt and 100 lbs. in a used one—a used belt being defined as one that has been run for more than 10 minutes. Knowing the correct tension and measuring it are two different matters. There is a belt-tension checking tool, but it rarely finds its way into even the most complete tool chest. If you happen to have one, by all means use it. Otherwise, use the *deflection method* for checking belt tension. Push firmly on the belt (about 10-lbs. pressure) and measure how much it deflects. A properly adjusted belt should deflect approximately 1/2 inch in the middle of an unsupported length of 14—18 inches. For new belts, the deflection can be slightly less, but after it's run for ten minutes or more, recheck it. If you want to be more accurate when using the deflection method, lay a straight edge on the backside of the belt to gage from.

Belt Adjustment—Depending on the year of your car and how it's equipped, there are three possible ways to adjust a drive belt. The most common, and the one you will have regardless of the year and how your car is equipped, is the type requiring that the accessory be rotated on a pivot. All alternators and rotary A/C compressors use this method as well as some power-steering pumps. To adjust a belt, rotate the accessory with a pry bar or large screwdriver to hold the belt in tension while you tighten the adjusting bolt to hold this tension. When doing this, make sure the pivot bolt is loose. And don't pry against something which is easily damaged such as a power-steering-pump reservoir. You won't have this problem with the rotary-type A/C compressor. There's a 1/2-inch square hole provided for using your 1/2-inch-drive breaker bar to tension its belt. It's a lot easier than trying to find something to pry against.

The second method is a variation on the method used to adjust the rotary A/C compressor, however this one came first. It's used for adjusting the two-piston-compressor belt. Rather than rotating the accessory, an adjustable idler pulley ro-

Fan shroud is laid back against the engine as the radiator is set in place.

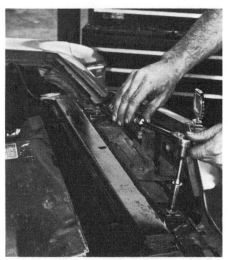
Cross-flow radiator in full-size Ford sits in a rubber-cushioned cradle and is captured at the top with this bracket which bolts to radiator support. Down-flow truck radiator is secured by bolting through its side flanges directly to the radiator support.

tates against the slack-side of the drive belt. Its adjustment is convenient because there is also a 1/2-inch-square hole provided in the pulley's bracket. The guy that came up with this one must have personally busted his knuckes a few times adjusting V-belts. Again, use a breaker bar to rotate the pulley against the belt, then tighten its adjusting bolt.

This method of adjusting belt tension is best: It's used for adjusting many of the mid-1970 and later power-steering-pump drive belts. Rather than rotating the pump on a pivot for belt adjustment, the pump uses two brackets that slide on each other. To adjust a belt using this type of arrangement, loosen the three clamping nuts, then rotate the nut on the adjusting stud which extends through the upper bracket. Turn the nut clockwise to tighten the belt. Be careful when doing this, it's easy to over-tighten a belt because of the little effort required to do the tightening as compared to the other methods. When you've reached the correct belt tension, torque the three clamping nuts 30—40 ft.lbs.

Radiator and Fan Shroud—Before installing the radiator, position the fan shroud back over the fan against the front of your engine. If you don't, you'll either have to remove the fan or the radiator again to install the shroud. Carefully lower the radiator into place and set it on the rubber pads if you have the cross-flow-type radiator, or start the mounting bolts through the radiator flange if yours is a down-flow radiator which mounts solidly to the radiator support. Secure the cross-flow-type radiator with its upper bracket.

Radiator installation completed by hookup of the transmission cooling lines and installation of new radiator hoses. Retighten hose clamps after running engine.

For the solid-mounted type, install its remaining bolts at the core's sides. Now you can slip out the cardboard you used for protecting your knuckles and the radiator core.

If your car is equipped with an automatic transmission, remove the hose you used to connect the transmission cooling lines, then connect the lines to the radiator fittings. Use a 5/8-inch tube-nut wrench to tighten the nuts. Don't over-tighten them—12 ft.lbs. maximum. You won't be able to use a torque wrench, however you should've developed a "feel" for torque by now. Install the upper and lower radiator hoses and you'll have the cooling system sealed. All it needs is coolant and the radiator cap. Complete the cooling system by installing the fan shroud. Set it in the Tinnerman clips at the bottom, and secure it with the two sheet-metal screws at the top if you have a cross-flow radiator, or use just screws in the case of a down-flow radiator.

After initially firing your engine—time it. Have your timing light ready and the distributor hold-down clamp loosened.

Two musts when installing an air-cleaner—a gasket under the air-cleaner and a clean filter. Another must is the fresh-air duct. Make sure yours is connected and in good condition. Duct tape is the best thing for making repairs.

After completing your preliminary engine running, drain the cooling system and recharge it with a good mix of anti-freeze. In addition to its anti-corrosive qualities it will also prolong water-pump-seal life and let your engine operate at a higher temperature before boiling.

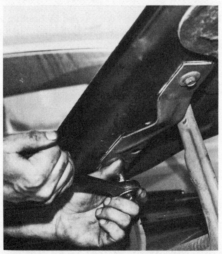
Tightening hood bolts after lining up hinges to marks on the hood. It should fit perfectly the first time, or as well as it did before removing it.

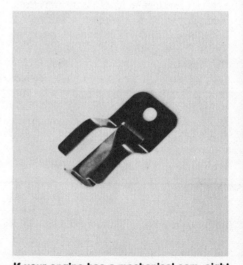
If your engine has a mechanical cam, eight rocker-arm clips like these from Mr. Gasket will deflect hot oil from your arms, clean engine and engine compartment while you set valve clearances. Buy them at a speed shop.

Engine Run-In Preparation—What you do prior to starting and running your newly rebuilt engine will ensure a long life for your engine. The first 30 minutes of running are the most critical for a new engine. The rules are, once the engine is fired, it should be kept running at *no less than 1500 RPM*, be well lubricated and cooled. Lubrication is particularly important to the camshaft. This is why the engine **should not be slow idled** under any circumstances during its initial running. Shut it down first.

If the cam lobes aren't drenched in oil during run-in, it's quite possible that some will be damaged. This means tearing your engine down, so I'm certain you don't want this situation. As for cooling, your car will be stationary, and consequently, not getting full air flow through the radiator. This is aggravated by the fact that a new engine generates more heat than one which is broken in, due to internal friction caused by tighter clearances. During run-in, you will notice that engine RPM will increase gradually as the engine runs, indicating that the engine is loosening up.

Before firing your engine, have two items on hand. If you have an automatic transmission and have drained its torque convertor, you'll need four quarts of ATF (automatic-transmission fluid) Type-F—Type CJ after 1976. Also, have your timing light hooked up so you can set timing immediately after it fires. Fluid is important because the front pump will fill the torque converter once the engine starts turning and the rest of the transmission will be low on fluid. So the first thing you should do once the engine fires is to add the ATF to prevent possible transmission damage.

To ensure your engine stays cool during its run-in, put your garden hose into service. Fill the radiator with water, then open the drain and adjust the water flow from the hose to the radiator to match what is coming out of the drain. This provides a source of cool water for your engine.

Engine Run-In—Before starting your engine, make sure it has oil pressure. Do this by disabling the ignition system, then crank the engine with the starter until oil pressure is indicated by the gage or idiot light. Within 15 seconds, the gage should indicate 35—40 psi with a standard oil pump, and up to 60 psi if you are using a high-volume pump. If you have an idiot light, all you'll know is the engine has oil pressure when the light goes out. Doing this will also help fill the carburetor float bowl/s with fuel. Have some fuel on hand to prime the carburetor after you have oil pressure and the ignition is reactivated.

Start your engine. Don't be alarmed by the puff of blue smoke and clattering lifters. They should both remedy themselves after the first few minutes of running unless you have a solid-liftered version of the 351C or 429 engine. Noisy lifters is one of the joys of these engines—to some. Adjust the idle speed if it's not right and keep your eye on the radiator. Add the automatic-transmission fluid and set the timing. When the thermostat opens, the water level will drop as water is drawn into the engine. Once you get the water level back up, maintain its level by adding water. Keep the engine running. Remember, *don't let it idle. Shut it down first.*

Make sure the distributor hold-down clamp is loose *before you start your engine*. Also, disconnect the distributor vacuum hose/s and plug them. Doubling the hose and taping it works well. Remember, rotating the distributor housing *clockwise advances timing; counter-clockwise retards* it. When you have the timing set, tighten the locking clamp and recheck the timing. Wait until after you shut the engine down to reconnect the vacuum hose/s. It's too dangerous to be fiddling around the front of the engine when it's running. Not only is there danger from a fan blade coming off—extremely remote, but it has happened—but also the fan is just waiting to grab a shirt tail, sleeve, cord or whatever to do serious bodily harm. The same thing goes for the accessory drive. **Be careful. Stay out of line with the fan.**

Post Run-In Checks—After initial run-in, shut the garden hose off and let the water drain out of the radiator. When it stops draining, shut the drain cock in preparation for refilling the radiator with permanent coolant. Check your engine from top to bottom for fluid leaks; gas, oil and water. If you spot any, remedy them now.

Because gaskets and hoses creep or relax when they are loaded and heated, you should go over the entire engine and retighten a few things. High on the list are the intake-manifold bolts, heater-hose and radiator-hose clamps. While you're at it, check the exhaust-manifold bolts too. The 351C Boss/HO head bolts should be retorqued to 125 ft.lbs. *hot.*

Engine Coolant—Regardless of the climate your car will be operated in, water should not be used as the sole coolant. If you use it as the main coolant, rust inhibitor must be added to prevent the interior of your engine from being damaged. I believe the best practice is to use anti-freeze so the coolant will have at least 0°F (-18°C) capability. This provides corrosion protection for your engine and also raises the temperature at which the coolant can operate before boiling.

Fill the radiator until it won't accept any more, then start the engine and wait for the thermostat to open. Have some anti-freeze handy. The engine will begin to purge itself of air, so you'll have to keep an eye on the coolant level. Don't forget the heater. To fill it, put the heater control on *heat* so its water valve will open if your heating system is so equipped. Coolant will then flow through the heater core and hoses. When the coolant level ceases to drop and is free of bubbles, cap the radiator. If you have a coolant-recovery system, fill the recovery bottle to the HOT LEVEL. Install a new radiator cap with the correct pressure rating.

Air Cleaner—You should now be ready to install the engine's crown, the air cleaner. Make sure there's a seal on the carburetor before installing the air-cleaner base. If it is not sealed, dust and dirt will enter the carburetor rather than being filtered out. If your engine is equipped with an exhaust-manifold heat duct or fresh-air duct, connect them. Install a new filter element and put the top on the filter. Secure it with a wing nut, but make sure there's a seal under the nut for the same reason as the seal between the air cleaner and carburetor. Connect any hoses or wires your air-cleaner assembly may require, such as the PCV hose.

Hood—You need to rustle up some help to install the hood. With the hood in place, run the attaching bolts in just short of snug and adjust the hood to the reference marks, or with the ice pick in the holes you drilled, and tighten the bolts. Close the hood to check its fit. It should be the same as before you removed it.

Trial Run—Now, with the satisfaction of having rebuilt your own engine, it's time to take it on its maiden voyage. Before you go tooling out of the driveway, make sure you've collected all your tools from the engine compartment. Your wrenches aren't going to do you any good distributed along the roadside. Take an inspection trip around your car to make sure you don't end up flattening your creeper, or whatever may have gotten left under your car in your haste to "see how she's gonna run."

It's not going to hurt your engine to take it up to normal highway speed on its first trip out, just avoid hard acceleration. Also, vary the speed. Drive just long enough to get your engine up to operating temperature and stabilized. Keep alert for any ominous sounds from under the hood or indications from your instrument panel. If something doesn't sound or look right, get off the road immediately and investigate.

After arriving back home, check fluid levels and look for any leaks that may have developed. Generally scrutinize the engine compartment. After everything appears to be OK, retorque the intake- and exhaust-manifold bolts, and check accessory drive-belt tension. You can now take a minute and admire *your engine* unless your engine is equipped with solid lifters. If so, you have one more duty to perform.

Hot-Lash Adjustment—If your engine is equipped with solid lifters, now is the time to do the *hot-lash* adjustment on the valves. It will be less than what you set them cold. This is because expansion of the valve-train components as they heat up changes clearances—and does so inconsistently. Therefore, lash must be reset with the valve train at its normal operating temperature. The cold setting was an approximation, but not a guarantee of ending up at the correct hot lash—0.025 inch for the 351C Boss and HO and 0.019 inch for the 429SCJ.

To adjust valves at their operating temperature, the engine should be running. This is a hot and messy job. To make it easier, acquire some *rocker-arm clips*. These fit over the pushrod end of the rocker arms and prevent oil from being thrown all over the side of your clean, newly rebuilt engine, engine compartment and you. Besides being messy, the oil is hot and smokes when it splashes on the hot exhaust manifolds. Rocker-arm clips are available at most hot-rod shops and are inexpensive.

If you have a 429SCJ *thread two of the valve-cover bolts* back in each head—the ones at the end of the top row. If you don't, you'll have such a vacuum leak in the end cylinders that they won't fire. Consequently, your engine won't idle so you can hot lash the valves.

Before adjusting the valves, warm up your engine to its normal operating temperature. After the engine is warmed up, shut it off and quickly remove the valve covers, being careful not to damage the gaskets. I find it's best to remove one cover at a time so you can do the valves in one head. Then replace the cover, warm the engine up again, and go through the same deal on the other side. Have your feeler gages and a 5/8-inch socket with a non-ratcheting handle ready. Remove all the jam nuts if you install them after hot lashing, and restart your engine. Work from one end of the head to the other with your gage and wrench. There should be a slight drag on the gage when the valve closes and the rocker arm unloads it. The valves will be opening and closing, so the feeler gage will be intermittently trapped between the valve and rocker arm from the time the lifter leaves the base circle until it's back on it. After going through all your valves, shut the engine off and remove the valve clips. Install the jam nuts while making sure the adjusting nuts don't move by holding them with another wrench. Replace the valve cover and go to the other side. When you finish, clean up your freshly oiled engine compartment, replace the ignition wires on the valve covers, and you're in business.

While putting the first 200 miles on your engine, constantly check the fluid levels—particularly oil—and be on guard for leaks. Keep checking things in the engine compartment in general. Correct any problems that might arise. Don't get excited if your engine uses a little oil. If you've used chrome rings, expect it to use oil for a while. Just keep the oil level up and change the oil and filter after the first 500 miles. Make these miles relatively easy ones and avoid operating your vehicle at sustained, or steady speeds. Vary speed as you drive. Avoid accelerating hard. Follow these rules and it will pay off in a longer lasting engine. After you've put 200—300 miles on your engine and you've readjusted its idle slower for what seems to be the last time, take your car or truck to a tune-up shop—one that specializes only in tune-ups—and have the engine expertly tuned. Now, your engine will not only be broken in, it will perform at its peak efficiency. For more information on tuning, go to the next chapter.

Retorque Head Bolts—As added insurance, retorque the head bolts using the same sequence and final torque figures quoted on page 115. This is particularly important with engines having compression in excess of 9.0:1. Because the valve covers have to come off for this, be careful when reinstalling to get the gaskets in place.

THANKS

The many versions of the 335 and 385 Series Ford engines and the large number of changes these engines have undergone over the years makes it impossible for any one person to write a book such as this. Completeness and accuracy required that I get help from people who make their living engineering Ford engines, selling or manufacturing engine parts, rebuilding engines and servicing them.

At the risk of leaving someone out, I am grateful to: Denny Wyckoff who was always ready to share his engine rebuilding experience. Daryl Koeppel of Jim Click Ford and Jim Hambacher and Steve Oathout of Holmes Tuttle Ford willingly took time to answer questions concerning parts, part numbers, interchangeability and those troublesome change levels. They also supplied many of the parts you see in photos throughout the book.

Don Wood provided words of wisdom about removing and installing engines, and tips on his favorite engine, the 429/460. Howard Aula, Allen Buchmaster, Charlie Camp, Guy Cramb, Bob Helvey, Mickey Matus and Ron Smaldone, all of Ford Motor Company, supplied much of the information in the Parts and Interchange chapter. Kevin Rotty and Bill Nelson opened up their tuneup shop for Saturday photo-shooting sessions.

Photos and drawings credited to Ford Motor company were cheerfully supplied by Linda Lee of Ford's Parts and Services Division. Others I would like to thank are: Ed Kerchen of Associated Spring, Bob Bub of Cloyes Gear and Products, Bob Lopez and John Thompson of Federal Mogul, Tom Tlusty of Muscle Parts, Cal DeBruin, Willey Thruman and Jeff Shumway of Sealed Power Corporation, Jack Little of Sun Electric, Terry Davis of TRW and Bill Borrusch and Bob Robertson.

10 | Tuneup

After putting a few hundred miles on your newly rebuilt engine, you should have it professionally tuned so it will provide you with peak performance. Even with the most complete tune-up set you're not going to be able to do much more than set dwell, initial timing, check point resistance, total advance and set your carburetor's fast idle, choke and idle mixture. This is particularly true with the later engines using electronic ignitions and complex emissions systems. Remember, you should not only be tuning for performance, but for minimum emissions. The biggest polluters on the roads are vehicles which have improperly tuned engines, not "old" engines. Typically, these cars do not perform as well as they could nor run as economically as they should.

PERFORMANCE AND EMISSIONS TUNING

The cost of a complete professional tune-up should range from $35 to $85, depending on the work required. The cost to you should be at the low end of the scale if you've done a proper job of rebuilding.

The type of tune-up equipment needed to tune your engine depends on how it's built and equipped. For instance, if you rebuilt your engine to original specifications, an electronic engine analyzer can be used. This will ensure your engine will perform *at least* as well as when it was new. However, if you modified your en-

351C Boss represents the ultimate in high-performance engine design from Detroit. Proper tuning is necessary for this or any other engine to operate at peak efficiency. Photo courtesy Ford.

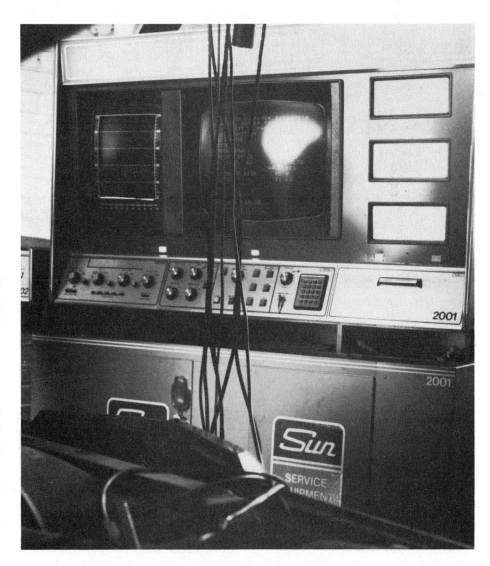

Having your engine professionally tuned is the best way of getting the most out of it. If your engine was rebuilt to original factory specifications, an electronic engine analyzer like this one will do the job. This Sun analyzer even has a printout so you'll have a record of your engine functions after the tune-up.

TUNEUP 157

Checking power where it counts at the rear wheels. Though increasing in use, few tune-up shops have chassis dynamometers. Even fewer have dynamometers capable of checking full horsepower, so check first if you want to check maximum output at the wheels.

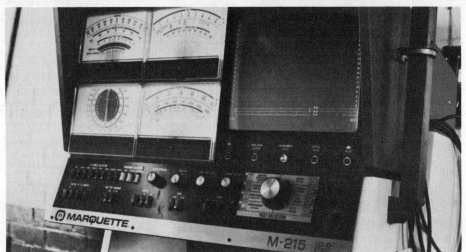
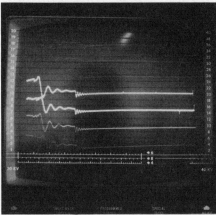

Electronic engine analyzer is capable of monitoring air/fuel ratio from the carburetor and checking distributor dwell at various engine RPM. Oscilloscope also monitors performance of each spark plug as shown in the scan.

gine such as installing a different camshaft, changing the carburetion system, ignition system or exhaust system, your engine must be recalibrated for it to obtain maximum performance. The only way of doing this is to tune it while its performance can be "seen" on a *chassis dynamometer*. The rear wheels of the car or truck drive *rolls* that are set flush in the garage floor. These rolls are adjusted to simulate the load the vehicle would apply to the road surface. Therefore, the chassis dynamometer permits an engine's functions to be monitored while it is operating under simulated driving conditions. In addition, the dynamometer can be used to determine the engine's power *at the drive wheels*. Consequently, it can be tuned for maximum power *and* mileage and minimum emissions. Beware of tuning for full power at full throttle only. This will compromise the more useful areas in the "performance curve." Tuned correctly, dividends in terms of power, mileage and reduced emissions will more than pay for the initial tune-up investment and ensure maximum pleasure and benefit from your newly rebuilt Ford engine.

Exhaust-gas analysis. Comparing hydrocarbon and carbon monoxide outputs is an actual situation method of determining whether an engine's fuel is being used efficiently. This device is valuable both when tuning for reduced emission levels and for economy—an increasingly important consideration.

OFFICIAL SHOP MANUALS

For additional information concerning your car or truck you should obtain the official Ford shop manual. Single-volume manuals covering specific vehicles were printed for trucks prior to 1966 and for cars prior to 1970. These single-volume manuals were superseded by five-volume manuals which cover *almost* all car or truck lines. They are grouped according to the vehicle system—chassis, engine, electrical and body. The fifth volume covers general maintenance and lubrication. These manuals are available from Helm. To order, or for additional information, contact Helm, Inc., P.O. Box 07150, Detroit, Michigan 48207. Telephone: (313) 865-5000. Make sure you include the model and year of your vehicle when ordering.

Index

A

A/C compressor 152—153
  back-side idler 147-148, 149
  bracket 16, 147
  clutch 144
  install 147-148
  remove 16
A-frame 12
ATF, see Automatic transmission, fluid
Accessory-drive belt
  adjust 152-153
  install 152-153
  remove 16
Air cleaner
  duct 55
  hose 155
  install 155
  wing nut 155
Air pump
  by-pass valve 148
  install 148
  remove 16
Alternator
  install 148
  remove 16, 20
Anti freeze 155
Anti-seize compound 110
Armature 124
  install 133
  remove 126
Automatic choke
  install 136
  remove 134-135
Automatic transmission
  bellhousing 19
  converter, see Converter
  cooling lines 14, 16, 19, 151, 153
  filler tube 16, 151
  fluid 137, 154
  front pump 137, 138
  front seal 137-138
  input shaft 21, 137, 138
  locating dowels 141, 142
  stator support 137, 138
  TV rod 15, 144

B

BDC 34, 46, 56, 105, 106, 117
Balance 25
Battery 13, 143
Battery cable
  connect 143
  disconnect 19
Bearing
  inspect 46, 47, 48
  edge ride 61
  spin 69
Bearing crush 69
Bearing scraper 92
Bellcrank 18
Bellhousing 19, 141, 142
  bolt pattern 25, 26-27
Blowby 3
Bore
  chamfer 58
  diameter 25-26, 29
  measure 53-56
  spacing 23
  taper 53, 55
  wear 53-55
Bottom-dead-center, see BDC
Breaker plate
  assemble 131, 132
  button 126, 131
  ground wire 131
  inspect 125
  install 131, 132
  remove 126
Breaker-plate pivot
  inspect 129, 130
Breaker points
  adjust 131-132, 133
  install 131
  remove 126
  rub block 131
Brinell test 27
Build-date code 22, 23

C

Cam bearing 51, 58
  bore 50
  chamfer 92
  install 91-94
  installation tool 50, 91
  journal 64-65
  oil hole 92, 93
  remove 49-50
  scrape 95
  set back 94
  size 91
  wear 50
Cam-bearing plug
  install 93, 94
  remove 50
Camshaft
  base circle 11, 64
  coating 95
  end-play 71
  inspection 64-65
  hydraulic 11, 74, 117-118
  install 94-95
  lubricate 94-95
  mechanical 9, 33, 74, 117
  plug 94
  remove 48
  runout 65
Camshaft lobe
  check 11
  design 64
  lift 6, 11, 48 64
  profile 64
  rake angle 64
  ramp 64
  toe 64
  wear 4, 6, 11, 94
Camshaft thrust plate
  install 95
  remove 48
Car wash 13
Carbon deposits 4, 6
Carburetor
  accelerator pump 135
  discharge nozzle 135
  EGR plate 123
  gasket 123, 136
  heat duct 40, 141
  identification tag 134
  idle screw 135, 136
  install 123
  power valve 135
  rebuild 134-136
  rebuild kit 134, 135, 136
  remove 39
  venturi booster 135
Carburetor float
  gage 135, 136
  install 135, 136
  level 135, 136
  remove 135
Carburetor linkage
  connect 144
  cable type 144, 145
  rod type 144, 145
  TV rod 144, 145
Carburetor spacer
  install 123
  remove 39
Cast iron
  grey 27
  nodular 27
Casting foundry 27
  Cleveland Foundry 27
  Michigan Casting Center 27
Casting number 22, 24
Centrifugal-advance weight 124
  install 129
  remove 127, 128
Chain hoist 12
Change level 22
Chassis dynamometer 3-4, 158
Cherry picker 12
Choke 7, 134,
Choke heat tube
  connect 123
  disconnect 39
Clearance volume 34-35
Clutch
  disc 138-139
  install 138-141
  linkage 16-17
  pressure plate 60, 139
  release bearing 141
  remove 39
Coil 16, 124, 144
Combustion chamber 6, 9, 34, 35-37
Combustion-chamber volume 27, 34, 35-37
Combustion leak 8-9, 51
Come-along 12
Compression ratio 27, 31, 34-35
Compression ring
  groove 68, 104
  install 104-105
  pip mark 104
  remove 66
Compression test 7-8
Condenser
  install 131
  remove 126
Connecting rod 26, 27, 30, 31, 69-71
  bearing bore 30, 69
  bolt 27, 30
  broached 30
  center-to-center length 30, 31
  number 45, 71, 105
  recondition 69-70
  side clearance 107
  spot faced 30
  wrist-pin bore 30, 70, 71
Connecting-rod bearing
  inspect 46, 47
  install 105
  remove 46
Connecting-rod-bearing bore
  out-of-round 69
  recondition 69-70
  taper 69
Connecting-rod-bearing cap
  grind 70
  install 106
  remove 45-46
Converter 20-21, 137
Converter housing 141, 142
  see also, Bellhousing
Coolant 9, 25, 40, 51, 155
Core plug
  install 93, 94
  remove 40
Crankshaft
  bearing-journal diameter 26, 60
  bearing-journal radius 61
  casting number 27
  counter-weight 27
  counterbalance 25
  end-play 102
  fatigue crack 61
  inspect 59-62
  install 95-102
  lubrication 59
  oil hole 62
  out-of-round 59, 60-61
  remove 46
  rear-main seal 62
  runout 57
  size 60
  taper 59, 61
  throw 27, 31
  thrust bearing 101
  thrust-face width 60
  351C 27
  351M/400 27
  429/460 27
Crankshaft bearing
  clearance 96-99
  inspect 46-49
  install 99
  remove 48
Crankshaft bearing-journal
  checking template 61
  polish 62
  radius 61
  surface finish 60, 61-62
Crankshaft damper
  install 110
  recondition 109
  remove 42-43
  repair sleeve 109
Crankshaft-damper spacer 109, 110
  install 110
  recondition 109
  remove 42-43
Crankshaft front seal 108-109
Crankshaft oil slinger
  damper 108
  install 108
  remove 43
Crankshaft thrust bearing
  seat 101-102
Creeper 18
Crossfire, see Induced current
Cross-hatch 53, 58
Cylinder block
  casting number 26, 27
  clean 51-53
  deck 56, 57
  deck clearance 31, 34
  deck height 23, 26, 27
  distributor bore 57
  inspect 51, 53-57
  notch 51
  rear face 27
  resurface 9
  sleeve 57
  water jacket 57
  weld 57
  351C 25-26
  351M/400 26-27
  429/460 27
Cylinder head 22, 31, 34-38
  assemble 88-89
  bolt 40, 114, 115
  clean 77
  crack 8, 77
  evolution 35, 37
  inspect 77-79
  install 114-115
  mill 34, 77-78
  notch 51
  remove 40-42
  resurface 9
  teardown 74-77
  warp 51, 77
  351C 35, 37
  351M/400 37
  429/460 37
Cylinder pressure 3, 4, 8, 9, 11

D

Deck, see Cylinder block, deck
Depth gage 10, 78
Detonation 5, 46, 66, 77
Dial bore-gage 53, 55
Dial indicator 9, 10, 11, 62, 102
Distributor
  assemble 129-133
  baseplate 126-128, 131-132
  breakerless 124
  bushing 129-129, 130
  cam 127, 128, 131
  cap 125, 126, 133, 148
  dual advance 124
  dual point 125, 131
  gear 127, 128, 129, 131
  inspect 129
  install 146-147
  magnetic pickup 124
  primary wire 126, 131, 133
  rebuild 126-133
  remove 39
  rotor 126, 148
  shaft 124, 125, 128, 129-130
  single point 126, 131
  sleeve 127, 128, 131
  time 146-147, 154
Dwell 125

E

EGR 123, 151
Engine
  code 22
  firing order 7, 11, 148
  install 37-56
  remove 12-21
  335 Series 23, 25
  385 Series 23, 25
Engine ground strap
  install 151
  remove 16
Engine hoist 137

Engine-identification
  decal 22
  tag 22
Engine mount
  bolt 18, 19, 141
  install 141, 142
  remove 39
Engine plate 141, 143
Engineering number 24
Equalizer bar, see clutch linkage
Exhaust manifold
  install 121
  O-ring 146
  remove 40
Exhaust pipe
  connect 143-144
  disconnect 18-19

F
Fan
  install 151-152
  remove 14
Fan shroud
  install 153
  remove 14
Flexplate 143
  install 141
  remove 39
Flywheel
  heat check 139-140
  hot spot 140-141
  inspect 138, 139
  install 141
  remove 39
Front cover 108, 110
  aluminum 108-109
  install 108, 110
  locating pin 110
  sheet metal 108, 110
  remove 43-44
Fuel line
  connect 144
  disconnect 18
  filter 136
Fuel pump
  install 122-123
  remove 39
Fuel-pump cam 107-108, 110
  install 107-108
  one piece 107
  remove 44
  two piece 107

G
Gage
  runout 83
  small-hole 78-79
  vacuum 6
Gasket adhesive 91, 119
Glaze breaking 56
Graph
  Bore Wear vs Ring End Gap 54
  Bore Taper vs Temperature 53
Ground strap 152

H
Head gasket 114
  blown 4, 8-9
  inspect 42
  install 114
  remove 42, 52
Heater hose
  disconnect 16
  install 150
Hood
  install 55
  remove 13
Hot tank 51, 58
Hydraulic lifter
  clean 65
  collapse 5
  inspect 65
  wear 5, 9
Hypalon sleeve 129-130

I
Idle speed 154, 155, 156
Induced current 150
Intake manifold
  baffle 118-119
  gasket 118-119
  install 118
  remove 40

J
Jack 137, 142
Jack stand 137

L
Lifter, also see types of
  bleed down 117
  clean 65
  design 64
  foot 63, 64
  inspect 64-65
  install 115
  prime 115
  remove 48-49
  wear 4, 6, 9-10, 11
Lifting lugs 141
  install 121
  remove 40
Loctite 108, 141

M
Magnetic pickup
  install 132-133
  remove 126, 127
Main bearing
  install 98-99
  remove 48
  size 95-96
Main bearing-journal
  check 96-97, 99
  clearance 96-97
Main seal, rear 95, 96
  install 99-100
  retaining pin 98, 99
  rope 96
  split lip 96
Main-bearing cap
  install 101
  register 101
  remove 46, 48
Manual transmission
  bearing retainer 137-138, 142
  bellhousing 19, 142
  front seal 137-138
  input shaft 142
Moly 110, 131, 141
Molybdenum-disulphide, see moly

O
Oil 44, 144
Oil consumption 5
Oil filter 144
Oil pan
  install 111-113
  gasket 112
  remove 43
  seal 111-112
Oil pump
  driveshaft 110-111
  install 111
  pickup 110
  remove 43
Oil-filter adapter
  install 94
  remove 50
Oil-gallery plug
  install 94
  remove 49
Oil-pressure sender 121-122, 144
Oil-pump driveshaft 43, 110-111, 129, 147
Oil-ring
  expander-spacer 104
  groove 68-69, 104
  install 103-104
  rail 104
  remove 66

P
PCV system 90
PVS (ported vacuum switch) 151
Part-number system 24
Pilot bearing 63
Pinging, see Detonation
Piston
  compression height 31, 32
  clean 67
  dish 31
  dome 31
  inspect 66-69
  knurl 56
  skirt 56, 66-67
  slap 5
Piston & connecting-rod assembly
  assemble 70-71
  disassemble 70-71
  install 105-106
  remove 45-46
Piston ring, see types of
  chrome 56, 57-58
  compress 105, 106
  end gap 103
  expander 103
  gapping 102, 103
  install 103-104
  moly 56, 57-58

plain cast-iron 56, 57-58
  twist 103
Piston-ring compressor 105, 106
Piston-to-bore clearance 55-56
Plastigage 91, 97, 99
Power-steering pump
  install 147-148, 149
  remove 16
Pressure plate
  heat check 139
  hot spot 139
  inspect 138, 139
  install 141
  remove 39
Preignition 4
Puller 42-43
Pushrod 115, 116
  guideplate 33
  hardened 33, 34
  inspect 116
  install 115
  length 34
  remove 40

R
Radiator
  cross flow 153
  down flow 153
  install 153
  recondition 14
  remove 13-14
Release bearing 141
Ridge reamer 45
Rocker arm
  adjustable 9, 33-34, 74-75, 115, 117
  baffle 41, 116, 117
  cast-iron 115, 116
  clip 155-156
  install 115
  non-adjustable 9, 33-34, 74-75, 115, 117-118
  pivot guided 32-33
  pushrod guided 33
  rail 31-32, 75
  remove 40
  stamped steel 115, 116
  valve-stem guided 31-32
Rocker-arm stud 9, 33, 37, 75, 78, 116
Rotor 124

S
Shop manual 158
Small-hole gage 78
Spark plug
  gap 141
  inspect 6
  install 115, 141, 146
Spark-plug lead
  clips 148, 150
  install 148, 150
  remove 7, 19
Spark-plug shield 121
Starter motor
  cable 19, 143
  install 142-143
  remove 19
Swept volume 34-35

T
TDC 9, 34, 106, 117, 146, 147
Table
  Cam-Drive Components 72
  Camshaft Bearing-Journal Specs. 65
  Camshaft Lift 10
  Compression Ratio 38
  Connecting-Rod Bore Dias. 70
  Crankshaft Casting Nos. 29
  Crankshaft Main-Bearing Dias. 95
  Crankshaft Specifications 60
  Cylinder-Block Casting Nos. 29
  Cylinder-Head Casting Nos. 28
  Cylinder-Head Specs. 37
  Deck Height & Clearance 32
  Engine Codes 23
  Head-Bolt Torque Specs. 115
  Intake-Manifold Milling 77
  Intake-Manifold Torque Specs. 120
  Pushrod Length 34
  Ring End-Gap vs Bore Wear 54
  Valve-Spring Specifications 86
  Vehicle Identification Codes 24
Tachometer 7
Tappet, see Lifter
Test
  compression 7-8
  leak-down 8
  power balance 6-7
  vacuum 6
Thermostat 41, 121
  bypass 23, 25, 26, 121

Throttle linkage 15, 144
Throttle shaft 134
Throttle solenoid 135, 136
Throwout bearing, see Release bearing
Thrust plate, see Camshaft thrust plate
Timing chain and sprocket
  heavy duty 71
  inspect 43-44
  install 106
  lubricate 113
  OEM 71
  remove 44-45
  roller 71
  wear 43-44
Timing-chain cover, see Front cover
Timing-chain sprocket 48
  install 106
  key 106
  remove 44-45
Timing gear, see Timing-chain sprocket
Timing light 5, 154
Tinnerman clip 153
Transmission, see type of
Tuneup 4, 157-158

U
UBS bolt 147

V
Vacuum diaphragm 124-125
  dual 125
  install 133
  remove 126
  single 125
Vacuum hose
  install 150-151
  remove 13, 16
Valve
  adjust 46-48, 116-117, 155-156
  burn 4, 42
  face 81, 82
  grind 81-82
  inspect 81
  install 87-89
  lap 83, 84
  lash 116, 117-118
  lift 9-11
  margin 82
  recondition 81-82
  seat, see Valve seat
  stem, see Valve stem
Valve cover 40, 156
Valve lifter, see Lifter
Valve guide
  insert 79-81
  knurl 79
  measure 78-79
  ream 79, 81
  wear 78-79
Valve seat
  angle 82, 83
  grind 82, 83
  runout 83
  width 83
Valve spring
  compressor 75-76, 88
  free height 84, 85, 87
  inspect 85-87
  install 87-88
  installed height 10-11, 84-85, 87-88
  load 84-86
  open height 10-11, 85
  rate 84
  remove 75-77
  retainer 74, 75-76, 88
  sag 85
  shim 86, 87, 88
  solid height
  squareness 85
Valve stem
  measure 81
  oversize 79
  seal 76, 88
  tip 74, 76, 81, 82
  wear 78, 79

W
Water pump 23
  install 109-110
  remove 39
Water-pump pulley
  install 151-152
  remove 16
Water-temperature sender 121-122, 144
Wiring harness, 16, 145
Woodruff key 106
Wrist pin
  noise 5
  install 70-71
  remove 70-71